STUDENT SOLUTIONS MANUAL

to accompany

MULTIVARIABLE CALCULUS

THIRD EDITION

William G. McCallum
University of Arizona

Deborah Hughes-Hallett
University of Arizona

Andrew M. Gleason
Harvard University

et al.

JOHN WILEY & SONS, INC.
New York • Chichester • Weinheim • Brisbane • Singapore • Toronto

COVER PHOTO © Eddie Hironaka/The Image Bank.

To order books or for customer service call 1-800-CALL-WILEY (225-5945).

ISBN 0-471-44193-7

Printed in the United States of America

10 9 8 7 6 5 4 3 2 1

Printed and bound by Victor Graphics, Inc.

CONTENTS

CHAPTER 12 193

CHAPTER 13 205

CHAPTER 14 217

CHAPTER 15 239

CHAPTER 16 257

CHAPTER 17 283

CHAPTER 18 299

CHAPTER 19 313

CHAPTER 20 321

APPENDIX 339

CHAPTER TWELVE

Solutions for Section 12.1

Exercises

1. The gravitational force on a 100 kg object which is $7,000,000$ meters from the center of the earth (or about 600 km above the earth's surface) is about 820 newtons.

5. The amount of money spent on beef equals the product of the unit price p and the quantity C of beef consumed:

$$M = pC = pf(I, p).$$

Thus, we multiply each entry in Table 12.1 on page 561 of the text by the price at the top of the column. This yields Table 12.1.

Table 12.1 *Amount of money spent on beef ($/household/week)*

Income \ Price	3.00	3.50	4.00	4.50
20	7.95	9.07	10.04	10.94
40	12.42	14.18	15.76	17.46
60	15.33	17.50	19.88	21.78
80	16.05	18.52	20.76	22.82
100	17.37	20.20	22.40	24.89

9. The distance of a point $P = (x, y, z)$ from the yz-plane is $|x|$, from the xz-plane is $|y|$, and from the xy-plane is $|z|$. So A is closest to the yz-plane, since it has the smallest x-coordinate in absolute value. B lies on the xz-plane, since its y-coordinate is 0. C is farthest from the xy-plane, since it has the largest z-coordinate in absolute value.

13. The distance formula: $d = \sqrt{(x_2 - x_1)^2 + (y_2 - y_1)^2 + (z_2 - z_1)^2}$ gives us the distance between any pair of points (x_1, y_1, z_1) and (x_2, y_2, z_2). Thus, we find

$$\text{Distance from } P_1 \text{ to } P_2 = 2\sqrt{2}$$
$$\text{Distance from } P_2 \text{ to } P_3 = \sqrt{6}$$
$$\text{Distance from } P_1 \text{ to } P_3 = \sqrt{10}$$

So P_2 and P_3 are closest to each other.

17. The graph is a plane parallel to the xz-plane, and passing through the point $(0, 1, 0)$. See Figure 12.1.

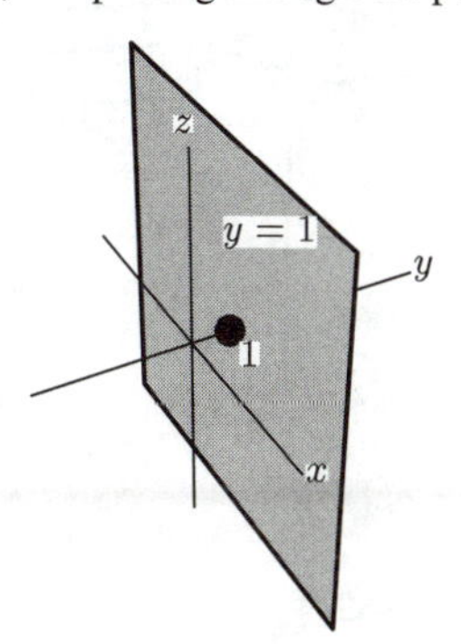

Figure 12.1

Problems

21. Each entry is the square of the y coordinate, so a possible formula is

$$f(x, y) = y^2.$$

25. (a) Holding x fixed at 4 means that we are considering an injection of 4 mg of the drug; letting t vary means we are watching the effect of this dose as time passes. Thus the function $f(4,t)$ describes the concentration of the drug in the blood resulting from a 4 mg injection as a function of time. Figure 12.2 shows the graph of $f(4,t) = te^{-t}$. Notice that the concentration in the blood from this dose is at a maximum at 1 hour after injection, and that the concentration in the blood eventually approaches zero.

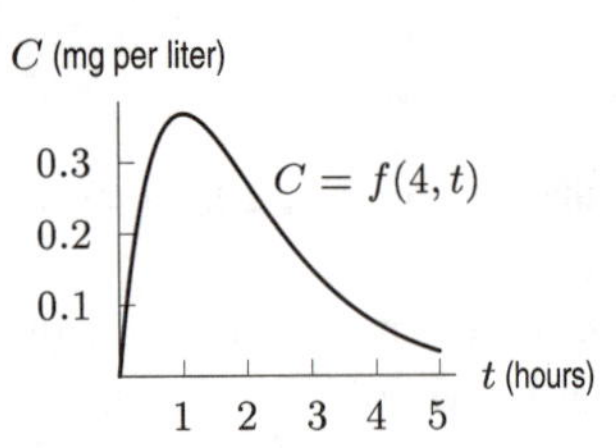

Figure 12.2: The function $f(4,t)$ shows the concentration in the blood resulting from a 4 mg injection

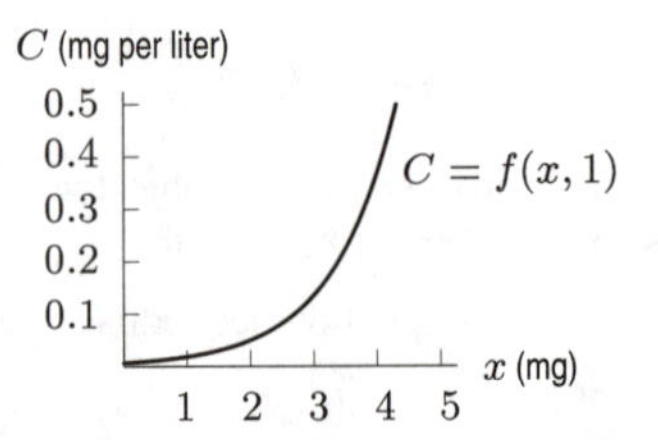

Figure 12.3: The function $f(x,1)$ shows the concentration in the blood 1 hour after the injection

(b) Holding t fixed at 1 means that we are focusing on the blood 1 hour after the injection; letting x vary means we are considering the effect of different doses at that instant. Thus, the function $f(x,1)$ gives the concentration of the drug in the blood 1 hour after injection as a function of the amount injected. Figure 12.3 shows the graph of $f(x,1) = e^{-(5-x)} = e^{x-5}$. Notice that $f(x,1)$ is an increasing function of x. This makes sense: If we administer more of the drug, the concentration in the bloodstream is higher.

29. By drawing the top four corners, we find that the length of the edge of the cube is 5. See Figure 12.4. We also notice that the edges of the cube are parallel to the coordinate axis. So the x-coordinate of the the center equals

$$-1 + \frac{5}{2} = 1.5.$$

The y-coordinate of the center equals

$$-2 + \frac{5}{2} = 0.5.$$

The z-coordinate of the center equals

$$2 - \frac{5}{2} = -0.5.$$

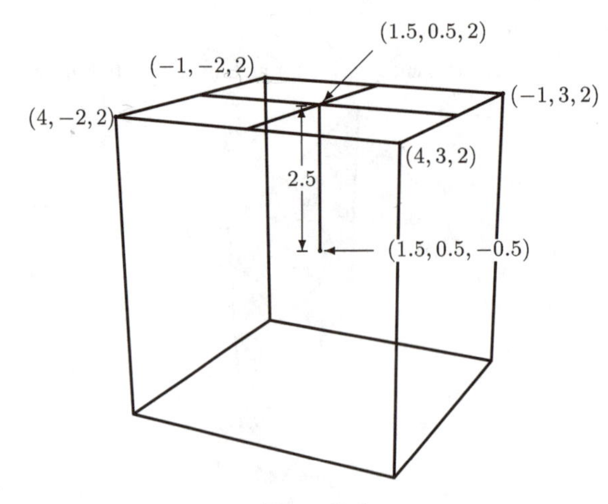

Figure 12.4

33. The distance from the y-axis is $d = \sqrt{x^2+z^2}$, so f is ruled out because it depends on y. Also, g takes different values at points the same distance from the y axis, for example, $g(1,0,0) = 0$ but $g(1/\sqrt{2}, 0, 1/\sqrt{2}) = 1/4$. So g is ruled out. On the other hand, $h(x,y,z) = 1/\sqrt{d^2+b^2}$, so (since b is a constant), h is a function of d alone.

Solutions for Section 12.2

Exercises

1. The graph is a horizontal plane 3 units above the xy-plane. See Figure 12.5.

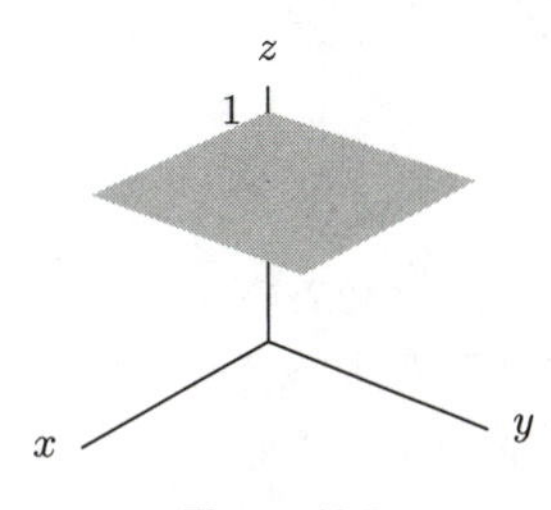

Figure 12.5

5. The graph is a plane with x-intercept 6, and y-intercept 3, and z-intercept 4. See Figure 12.6.

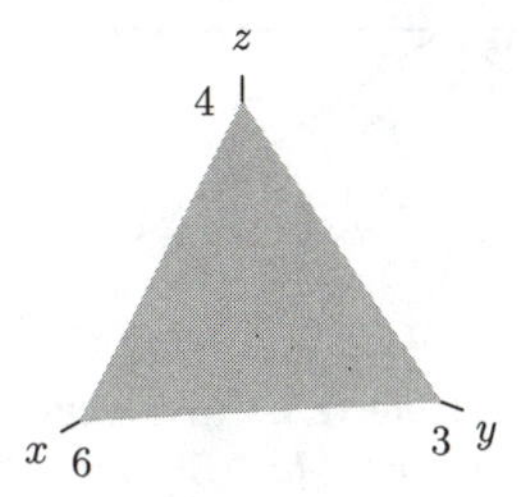

Figure 12.6

9. **(a)** The value of z only depends on the distance from the point (x, y) to the origin. Therefore the graph has a circular symmetry around the z-axis. There are two such graphs among those depicted in the figure in the text: I and V. The one corresponding to $z = \frac{1}{x^2+y^2}$ is I since the function blows up as (x, y) gets close to $(0, 0)$.

(b) For similar reasons as in part (a), the graph is circularly symmetric about the z-axis, hence the corresponding one must be V.

(c) The graph has to be a plane, hence IV.

(d) The function is independent of x, hence the corresponding graph can only be II. Notice that the cross-sections of this graph parallel to the yz-plane are parabolas, which is a confirmation of the result.

(e) The graph of this function is depicted in III. The picture shows the cross-sections parallel to the zx-plane, which have the shape of the cubic curves $z = x^3 -$ constant.

Problems

13. **(a)** This is a bowl; z increases as the distance from the origin increases, from a minimum of 0 at $x = y = 0$.

(b) Neither. This is an upside-down bowl. This function will decrease from 1, at $x = y = 0$, to arbitrarily large negative values as x and y increase due to the negative squared terms of x and y. It will look like the bowl in part (a) except flipped over and raised up slightly.

(c) This is a plate. Solving the equation for z gives $z = 1 - x - y$ which describes a plane whose x and y slopes are -1. It is perfectly flat, but not horizontal.

(d) Within its domain, this function is a bowl. It is undefined at points at which $x^2 + y^2 > 5$, but within those limits it describes the bottom half of a sphere of radius $\sqrt{5}$ centered at the origin.

(e) This function is a plate. It is perfectly flat and horizontal.

17. **(a)** Cross-sections with x fixed at $x = b$ are in Figure 12.7.

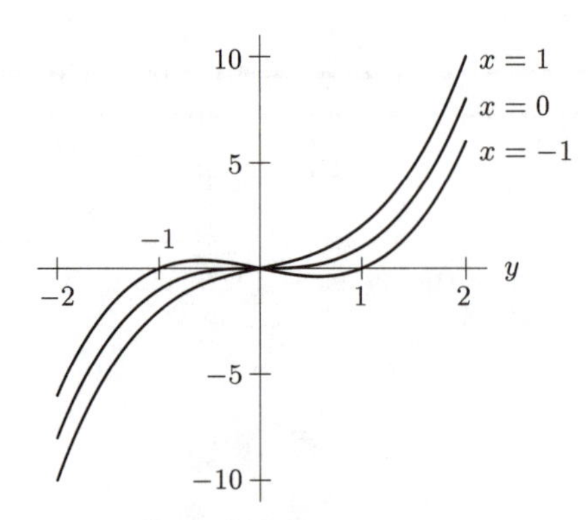

Figure 12.7: Cross-section
$f(a, y) = y^3 + ay$, with $a = -1,\ 0,\ 1$

(b) Cross-section with y fixed at $y = 6$ are in Figure 12.8.

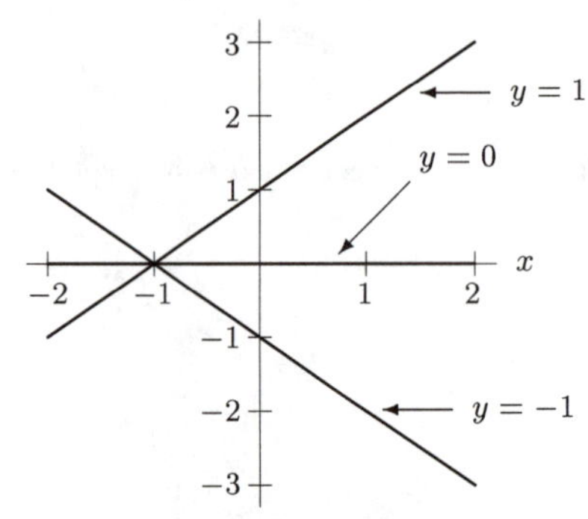

Figure 12.8: Cross-section
$f(x, b) = b^3 + bx$, with $b = -1,\ 0,\ 1$

21. **(a)** The plane $y = 1$ intersects the graph in the parabola $z = (x^2 + 1)\sin(1) + x = x^2\sin(1) + x + \sin(1)$. Since $\sin(1)$ is a constant, $z = x^2\sin(1) + x + \sin(1)$ is a quadratic function whose graph is a parabola.

Any plane of the form $y = a$ will do as long as a is not a multiple of π.

(b) The plane $y = \pi$ intersects the graph in the straight line $z = \pi^2 x$. (Since $\sin\pi = 0$, the equation becomes linear, $z = \pi^2 x$ if $y = \pi$.)

(c) The plane $x = 0$ intersects the graph in the curve $z = \sin y$.

Solutions for Section 12.3

Exercises

1. The contour where $f(x, y) = x + y = c$, or $y = -x + c$, is the graph of the straight line with slope -1 as shown in Figure 12.9. Note that we have plotted the contours for $c = -3, -2, -1, 0, 1, 2, 3$.

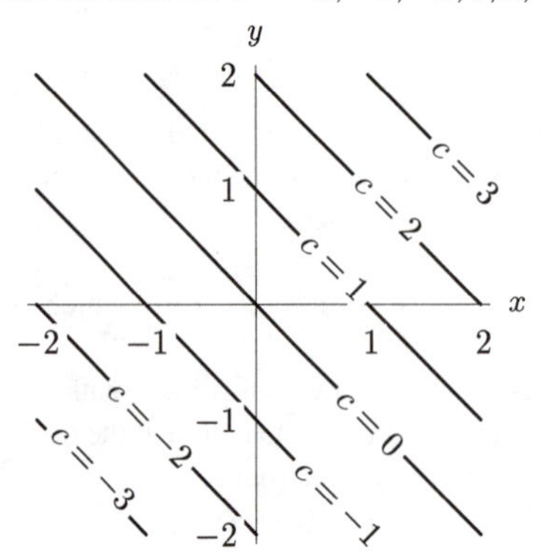

Figure 12.9

5. The contour where $f(x,y) = -x^2 - y^2 + 1 = c$, where $c \leq 1$, is the graph of the circle centered at $(0,0)$, with radius $\sqrt{1-c}$ as shown in Figure 12.10. Note that we have plotted the contours for $c = -3, -2, -1, 0, 1$. The contours become more closely packed as we move further from the origin.

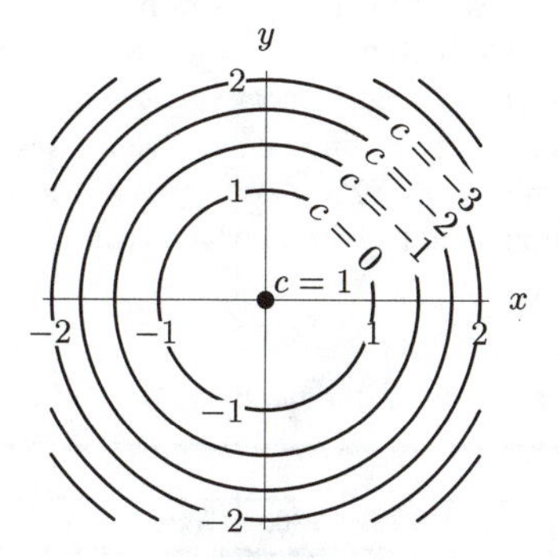

Figure 12.10

9. The contour where $f(x,y) = \cos(\sqrt{x^2+y^2}) = c$, where $-1 \leq c \leq 1$, is a set of circles centered at $(0,0)$, with radius $\cos^{-1} c + 2k\pi$ with $k = 0, 1, 2, ..$ and $-\cos^{-1} c + 2k\pi$, with $k = 1, 2, 3, ...$ as shown in Figure 12.11. Note that we have plotted contours for $c = 0, 0.2, 0.4, 0.6, 0.8, 1$.

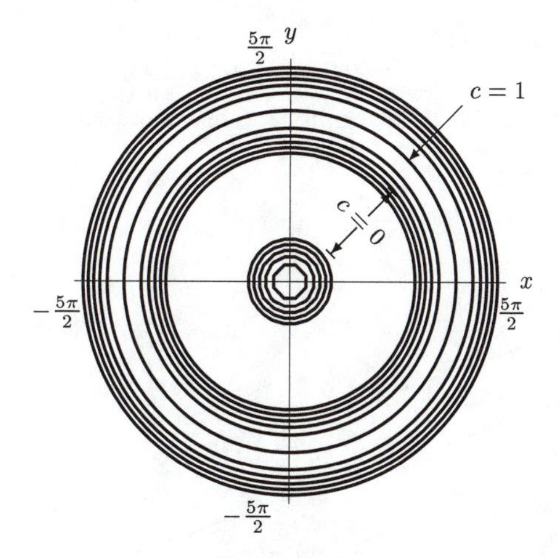

Figure 12.11

13. See Figure 12.12.

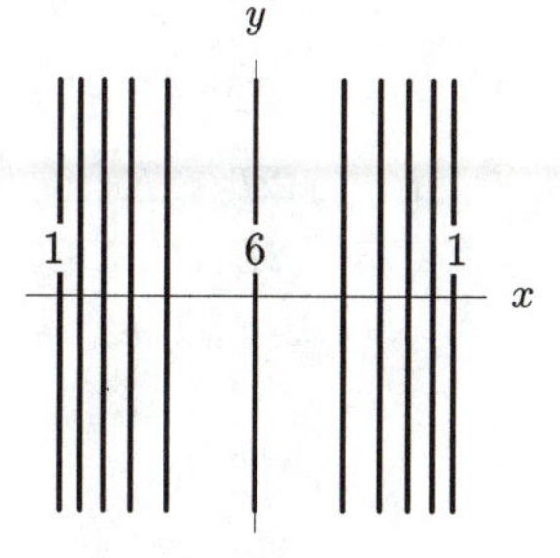

Figure 12.12

Problems

17. **(a)** The point representing 13% and \$6000 on the graph lies between the 120 and 140 contours. We estimate the monthly payment to be about \$137.

(b) Since the interest rate has dropped, we will be able to borrow more money and still make a monthly payment of \$137. To find out how much we can afford to borrow, we find where the interest rate of 11% intersects the \$137 contour and read off the loan amount to which these values correspond. Since the \$137 contour is not shown, we estimate its position from the \$120 and \$140 contours. We find that we can borrow an amount of money that is more than \$6000 but less than \$6500. So we can borrow about \$250 more without increasing the monthly payment.

(c) The entries in the table will be the amount of loan at which each interest rate intersects the 137 contour. Using the \$137 contour from (b) we make table 12.2.

Table 12.2 *Amount borrowed at a monthly payment of \$137.*

Interest Rate (%)	0	1	2	3	4	5	6	7
Loan Amount (\$)	8200	8000	7800	7600	7400	7200	7000	6800
Interest rate (%)	8	9	10	11	12	13	14	15
Loan Amount (\$)	6650	6500	6350	6250	6100	6000	5900	5800

21. **(a)** The profit is given by the following:

$$\pi = \text{Revenue from } q_1 + \text{ Revenue from } q_2 - \text{Cost}.$$

Measuring π in thousands, we obtain:

$$\pi = 3q_1 + 12q_2 - 4.$$

(b) A contour diagram of π follows. Note that the units of π are in thousands.

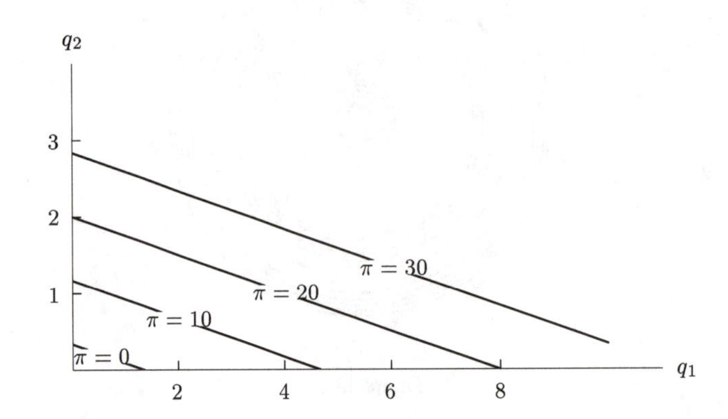

25. Suppose P_0 is the production given by L_0 and K_0, so that

$$P_0 = f(L_0, K_0) = cL_0^\alpha K_0^\beta.$$

We want to know what happens to production if L_0 is increased to $2L_0$ and K_0 is increased to $2K_0$:

$$\begin{aligned} P &= f(2L_0, 2K_0) \\ &= c(2L_0)^\alpha (2K_0)^\beta \\ &= c2^\alpha L_0^\alpha 2^\beta K_0^\beta \\ &= 2^{\alpha+\beta} cL_0^\alpha K_0^\beta \\ &= 2^{\alpha+\beta} P_0. \end{aligned}$$

Thus, doubling L and K has the effect of multiplying P by $2^{\alpha+\beta}$. Notice that if $\alpha+\beta > 1$, then $2^{\alpha+\beta} > 2$, if $\alpha+\beta = 1$, then $2^{\alpha+\beta} = 2$, and if $\alpha + \beta < 1$, then $2^{\alpha+\beta} < 2$. Thus, $\alpha + \beta > 1$ gives increasing returns to scale, $\alpha + \beta = 1$ gives constant returns to scale, and $\alpha + \beta < 1$ gives decreasing returns to scale.

29. **(a)** See Figure 12.13.

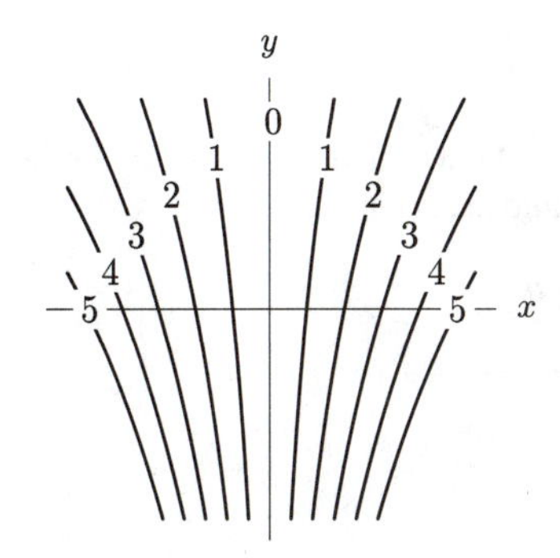

Figure 12.13

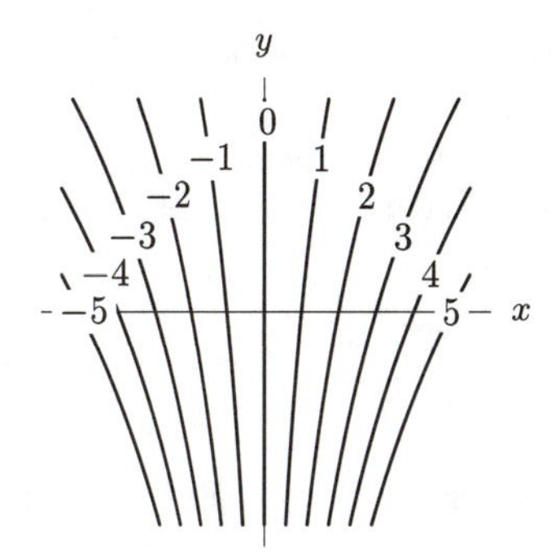

Figure 12.14

(b) See Figure 12.14.

33. **(a)**

(i)

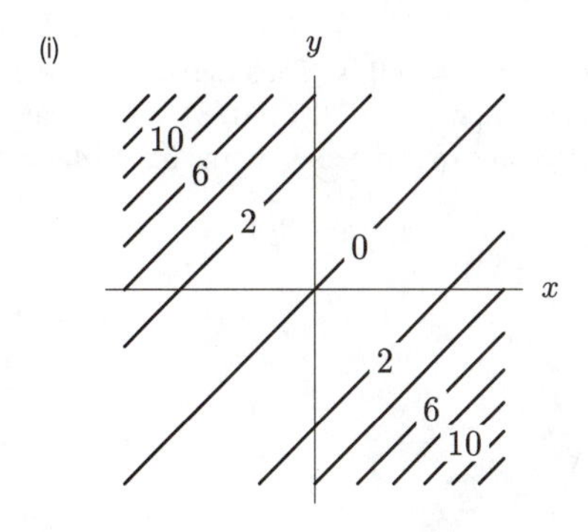

(ii)

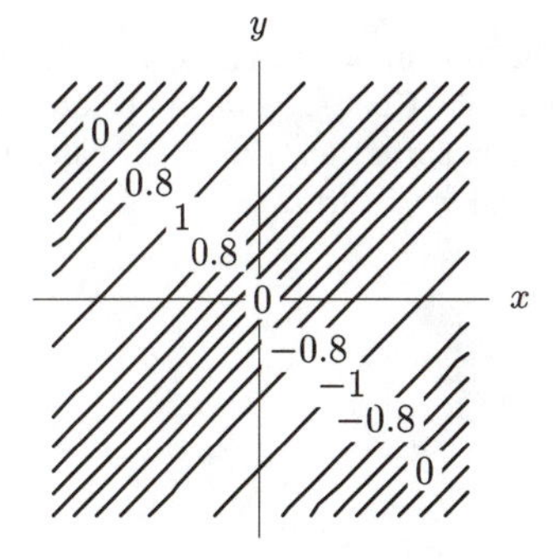

(b) The function $f(x,y) = g(y-x)$ is constant on lines $y - x = k$. Thus all lines parallel to $y = x$ are level curves of f.

Solutions for Section 12.4

Exercises

1. A table of values is linear if the rows are all linear and have the same slope and the columns are all linear and have the same slope. We see that the table might represent a linear function since the slope in each row is 3 and the slope in each column is -4.

5. **(a)** Yes.

(b) The coefficient of m is 15 dollars per month. It represents the monthly charge to use this service. The coefficient of t is 0.05 dollars per minute. Each minute the customer is on-line costs 5 cents.

(c) The intercept represents the base charge. It costs \$35 just to get hooked up to this service.

(d) We have $f(3, 800) = 120$. A customer who uses this service for three months and is on-line for a total of 800 minutes is charged \$120.

9. A contour diagram is linear if the contours are parallel straight lines, equally spaced for equally spaced values of z. This contour diagram could represent a linear function.

13. In the diagram the contours correspond to values of the function that are 15 units apart, i.e., there are contours for $-90, -75, -60$, etc. An increase of 3 units in the y direction moves you from one contour to the next and changes the function by -15, so the y slope is $-15/3 = -5$. Similarly, an increase of 6 in the x direction crosses two contour lines and changes the function by 30; so the x slope is $30/6 = 5$. Hence $f(x,y) = c + 5x - 5y$. We see from the diagram that $f(8,4) = -75$. Solving for c gives $c = -95$. Therefore the function is $f(x,y) = -95 + 5x - 5y$

Problems

17. Let the equation of the plane be

$$z = c + mx + ny$$

Since we know the points: $(4, 0, 0)$, $(0, 3, 0)$, and $(0, 0, 2)$ are all on the plane, we know that they satisfy the same equation. We can use these values of (x, y, z) to find c, m, and n. Putting these points into the equation we get:

$$0 = c + m \cdot 4 + n \cdot 0 \quad \text{so } c = -4m$$

$$0 = c + m \cdot 0 + n \cdot 3 \quad \text{so } c = -3n$$

$$2 = c + m \cdot 0 + n \cdot 0 \quad \text{so } c = 2$$

Because we have a value for c, we can solve for m and n to get

$$c = 2, m = -\frac{1}{2}, n = -\frac{2}{3}.$$

So the linear function is

$$f(x, y) = 2 - \frac{1}{2}x - \frac{2}{3}y.$$

21. The time in minutes to go 10 miles at a speed of s mph is $(10/s)(60) = 600/s$. Thus the 120 lb person going 10 mph uses $(7.4)(600/10) = 444$ calories, and the 180 lb person going 8 mph uses $(7.0)(600/8) = 525$ calories. The 120 lb person burns $444/120 = 3.7$ calories per pound for the trip, while the 180 lb person burns $525/180 = 2.9$ calories per pound for the trip.

25.

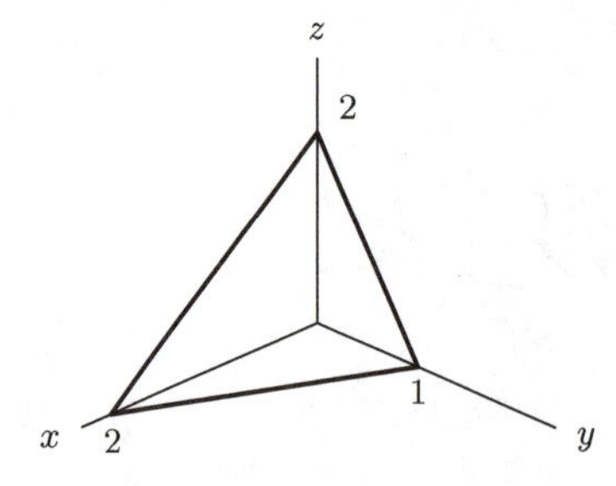

29. **(a)** Expenditure, E, is given by the equation:

$$E = (\text{price of raw material } 1)m_1 + (\text{price of raw material } 2)m_2 + C$$

where C denotes all the other expenses (assumed to be constant). Since the prices of the raw materials are constant, but m_1 and m_2 are variables, we have a linear function.

(b) Revenue, R, is given by the equation:

$$R = (p_1)q_1 + (p_2)q_2.$$

Since p_1 and p_2 are constant, while q_1 and q_2 are variables, we again have a linear function.

(c) Revenue is again given by the equation,

$$R = (p_1)q_1 + (p_2)q_2.$$

Since p_2 and q_2 are now constant, the term $(p_2)q_2$ is also constant. However, since p_1 and q_1 are variables, the $(p_1)q_1$ term means that the function is not linear.

Solutions for Section 12.5

Exercises

1. **(a)** Observe that setting $f(x, y, z) = c$ gives a cylinder about the x-axis, with radius $\sqrt{c}$. These surfaces are in graph (I).

(b) By the same reasoning the level curves for $h(x, y, z)$ are cylinders about the y-axis, so they are represented in graph (II).

5. The bottom half of the ellipsoid is represented by

$$z = f(x, y) = -\sqrt{2(1 - x^2 - y^2)}$$
$$g(x, y, z) = x^2 + y^2 + \frac{z^2}{2} = 1.$$

Other answers are possible

9. No, because $z = \sqrt{x^2 + 3y^2}$ and $z = -\sqrt{x^2 + 3y^2}$, so some z-values correspond to two points on the surface.

13. If we solve for z, we get $z = (1 - x^2 - y)^2$, so the level surface is the graph of $f(x, y) = (1 - x^2 - y)^2$.

17. An elliptic paraboloid.

Problems

21. In the xz-plane, the equation $x^2/4 + z^2 = 1$ is an ellipse, with widest points at $x = \pm 2$ on the x-axis and crossing the z-axis at $z = \pm 1$. Since the equation has no y term, the level surface is a cylinder of elliptical cross-section, centered along the y-axis.

25. The level surfaces are the graphs of $\sin(x + y + z) = k$ for constant k (with $-1 \le k \le 1$). This means $x + y + z = \sin^{-1}(k) + 2\pi n$, or $\pi - \sin^{-1}(k) + 2n\pi$ for all integers n. Therefore for each value of k, with $-1 \le k \le 1$, we get an infinite family of parallel planes. So the level surfaces are families of parallel planes.

29. Starting with the equation $z = \sqrt{x^2 + y^2}$, we flip the cone and shift it up one, yielding $z = 1 - \sqrt{x^2 + y^2}$. This is a cone with vertex at $(0, 0, 1)$ that intersects the xy-plane in a circle of radius 1. Interchanging the variables, we see that $y = 1 - \sqrt{x^2 + z^2}$ is an equation whose graph includes the desired cone C. Finally, we express this equation as a level surface $g(x, y, z) = 1 - \sqrt{x^2 + z^2} - y = 0$.

Solutions for Section 12.6

Exercises

1. No, $1/(x^2 + y^2)$ is not defined at the origin, so is not continuous at all points in the square $-1 \le x \le 1, -1 \le y \le 1$.

5. The function $\tan(\theta)$ is undefined when $\theta = \pi/2 \approx 1.57$. Since there are points in the square $-2 \le x \le 2, -2 \le y \le 2$ with $x \cdot y = \pi/2$ (e.g. $x = 1, y = \pi/2$) the function $\tan(xy)$ is not defined inside the square, hence not continuous.

9. Since f doesn't depend on y we have:

$$\lim_{(x,y)\to(0,0)} f(x, y) = \lim_{x\to 0} \frac{x}{x^2 + 1} = \frac{0}{0 + 1} = 0.$$

Problems

13. Points along the positive x-axis are of the form $(x, 0)$; at these points the function looks like $x/2x = 1/2$ everywhere (except at the origin, where it is undefined). On the other hand, along the y-axis, the function looks like $y^2/y = y$, which approaches 0 as we get closer to the origin. Since approaching the origin along two different paths yields numbers that aren't the same, the limit doesn't exist.

17. **(a)** We have $f(x, 0) = 0$ for all x and $f(0, y) = 0$ for all y, so these are both continuous (constant) functions of one variable.

(b) The contour diagram suggests that the contours of f are lines through the origin. Providing it is not vertical, the equation of such a line is

$$y = mx.$$

To confirm that such lines are contours of f, we must show that f is constant along these lines. Substituting into the function, we get

$$f(x, y) = f(x, mx) = \frac{x(mx)}{x^2 + (mx)^2} = \frac{mx^2}{x^2 + m^2x^2} = \frac{m}{1 + m^2} = \text{constant}.$$

Since $f(x, y)$ is constant along the line $y = mx$, such lines are contained in contours of f.

(c) We consider the limit of $f(x, y)$ as $(x, y) \to (0, 0)$ along the line $y = mx$. We can see that

$$\lim_{x \to 0} f(x, mx) = \frac{m}{1 + m^2}.$$

Therefore, if $m = 1$ we have

$$\lim_{\substack{(x,y)\to(0,0)\\ y=x}} f(x, y) = \frac{1}{2}$$

whereas if $m = 0$ we have

$$\lim_{\substack{(x,y)\to(0,0)\\ y=0}} f(x, y) = 0.$$

Thus, no matter how close we are to the origin, we can find points (x, y) where the value $f(x, y)$ is $1/2$ and points (x, y) where the value $f(x, y)$ is 0. So the limit $\lim_{(x,y)\to(0,0)} f(x, y)$ does not exist. Thus, f is not continuous at $(0, 0)$, even though the one-variable functions $f(x, 0)$ and $f(0, y)$ are continuous at $(0, 0)$. See Figures 17 and 17

21. The function, f is continuous at all points (x, y) with $x \neq 3$. We analyze the continuity of f at the point $(3, a)$. We have:

$$\lim_{(x,y)\to(3,a), x<3} f(x, y) = \lim_{y\to a} (c + y) = c + a$$

$$\lim_{(x,y)\to(3,a), x>3} f(x, y) = \lim_{x>3, x\to 3} (5 - x) = 2.$$

We want to see if we can find one value of c such that $c + a = 2$ for all a. This would mean that $c = 2 - a$, but then c would be dependent on a. Therefore, we cannot make the function continuous everywhere.

Solutions for Chapter 12 Review

Exercises

1. These conditions describe a line parallel to the z-axis which passes through the xy-plane at $(2, 1, 0)$.

5. Contours are lines of the form $3x - 5y + 1 = c$ as shown in Figure 12.15. Note that for the regions of x and y given, the c values range from $-12 < c < 12$ and are equally spaced by 4.

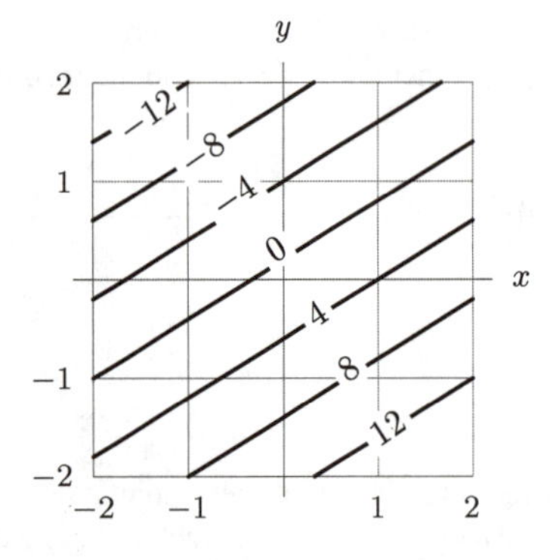

Figure 12.15

9. The function h decreases as y increases: each increase of y by 2 takes you down one contour and hence changes the function by 2, so the slope in the y direction is -1. The slope in the x direction is 2, so the formula is $h(x, y) = c + 2x - y$. From the diagram we see that $h(0, 0) = 4$, so $c = 4$. Therefore, the formula for this linear function is $h(x, y) = 4 + 2x - y$.

13. The paraboloid is $z = x^2 + y^2 + 5$, so it is represented by

$$z = f(x, y) = x^2 + y^2 + 5$$

and

$$g(x, y, z) = x^2 + y^2 + 5 - z = 0.$$

Other answers are possible.

Problems

17. Might be true. The function $z = x^2 - y^2 + 1$ has this property. The level curve $z = 1$ is the lines $y = x$ and $y = -x$.

21. We complete the square

$$x^2 + 4x + y^2 - 6y + z^2 + 12z = 0$$
$$x^2 + 4x + 4 + y^2 - 6y + 9 + z^2 + 12z + 36 = 4 + 9 + 36$$
$$(x+2)^2 + (y-3)^2 + (z+6)^2 = 49$$

The center is $(-2, 3, -6)$ and the radius is 7.

25. The level surfaces have equation $\cos(x+y+z) = c$. For each value of c between -1 and 1, the level surface is an infinite family of planes parallel to $x + y + z = \arccos(c)$. For example, the level surface $\cos(x + y + z) = 0$ is the family of planes

$$x + y + z = \frac{\pi}{2} \pm 2n\pi, \quad n = 0, 1, 2, \ldots.$$

29. One possible equation: $z = (x - y)^2$. See Figure 12.16.

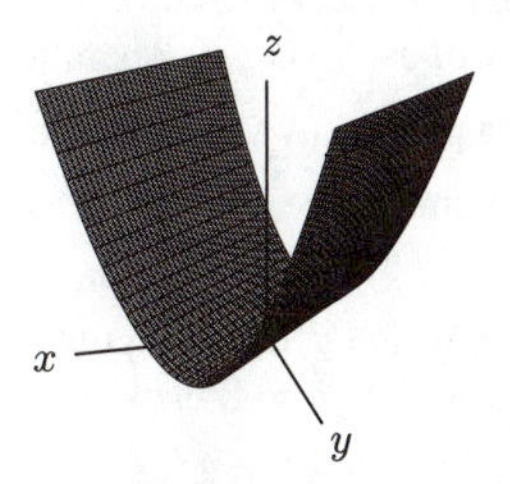

Figure 12.16

33. (a) For $g(x, t) = \cos 2t \sin x$, our snapshots for fixed values of t are still one arch of the sine curve. The amplitudes, which are governed by the $\cos 2t$ factor, now change twice as fast as before. That is, the string is vibrating twice as fast.

(b) For $y = h(x, t) = \cos t \sin 2x$, the vibration of the string is more complicated. If we hold t fixed at any value, the snapshot now shows one full period, i.e. one crest and one trough, of the sine curve. The magnitude of the sine curve is time dependent, given by $\cos t$. Now the center of the string, $x = \pi/2$, remains stationary just like the end points. This is a vibrating string with the center held fixed, as shown in Figure 12.17.

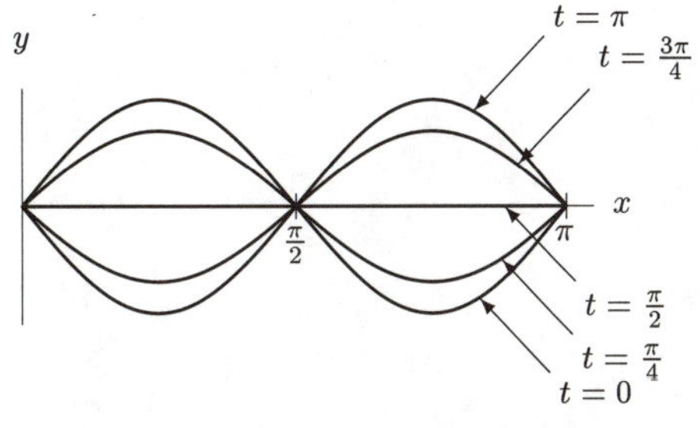

Figure 12.17: Another vibrating string: $y = h(x, t) = \cos t \sin 2x$

CHECK YOUR UNDERSTANDING

1. True. Since each choice of x and y determines a unique value for $f(x, y)$, choosing $x = 10$ yields a unique value of $f(10, y)$ for any choice of y.

5. False. Fixing $w = k$ gives the one-variable function $g(v) = e^v/k$, which is an increasing exponential function if $k > 0$, but is decreasing if $k < 0$.

9. True. For example, consider the weekly beef consumption C of a household as a function of total income I and the cost of beef per pound p. It is possible that consumption increases as income increases (for fixed p) and consumption decreases as the price of beef increases (for fixed I).

13. False. The plane $z = 2$ is parallel to the xy-plane.

17. True. The origin is the closest point in the yz-plane to the point $(3, 0, 0)$, and its distance to $(3, 0, 0)$ is 3.

21. True. The cross-section with $y = 1$ is the line $z = x + 1$.

25. True. The intersection, where $f(x, y) = g(x, y)$, is given by $x^2 + y^2 = 1 - x^2 - y^2$, or $x^2 + y^2 = 1/2$. This is a circle of radius $1/\sqrt{2}$ parallel to the xy-plane at height $z = 1/2$.

29. False. For example, the y-axis intersects the graph of $f(x, y) = 1 - x^2 - y^2$ twice, at $y = \pm 1$.

33. False. The graph could be a hemisphere, a bowl-shape, or any surface formed by rotating a curve about a vertical line.

37. False. The fact that the $f = 10$ and $g = 10$ contours are identical only says that one horizontal slice through each graph is the same, but does not imply that the entire graphs are the same. A counterexample is given by $f(x, y) = x^2 + y^2$ and $g(x, y) = 20 - x^2 - y^2$.

41. False. Any two-variable function that is missing one variable (e.g. $f(x, y) = x^2$) will have parallel lines for contours. Linear functions have the additional property of *evenly-spaced* parallel lines for contours.

45. True. Since the graph of a linear function is a plane, any vertical slice parallel to the yz-plane will yield a line.

49. False. All of the columns have to have the same slope, as do the rows, but the row slopes can differ from the column slopes.

53. True. The graph of $f(x, y)$ is the set of all points (x, y, z) satisfying $z = f(x, y)$. If we define the three-variable function g by $g(x, y, z) = f(x, y) - z$, then the level surface $g = 0$ is exactly the same as the graph of $f(x, y)$.

57. False. The level surface $g = 0$ of the function $g(x, y, z) = x^2 + y^2 + z^2$ consists of only the origin.

CHAPTER THIRTEEN

Solutions for Section 13.1

Exercises

1. $\|\vec{z}\| = \sqrt{(1)^2 + (-3)^2 + (-1)^2} = \sqrt{1+9+1} = \sqrt{11}$.

5. $\|\vec{y}\| = \sqrt{(4)^2 + (-7)^2} = \sqrt{16+49} = \sqrt{65}$.

9. $-4\vec{i} + 8\vec{j} - 0.5\vec{i} + 0.5\vec{k} = -4.5\vec{i} + 8\vec{j} + 0.5\vec{k}$

13. $\|\vec{v}\| = \sqrt{1.2^2 + (-3.6)^2 + 4.1^2} = \sqrt{31.21} \approx 5.6$.

17. **(a)** See Figure 13.1.
(b) $\|\vec{v}\| = \sqrt{5^2 + 7^2} = \sqrt{74} = 8.602$.
(c) We see in Figure 13.2 that $\tan\theta = \frac{7}{5}$ and so $\theta = 54.46^\circ$.

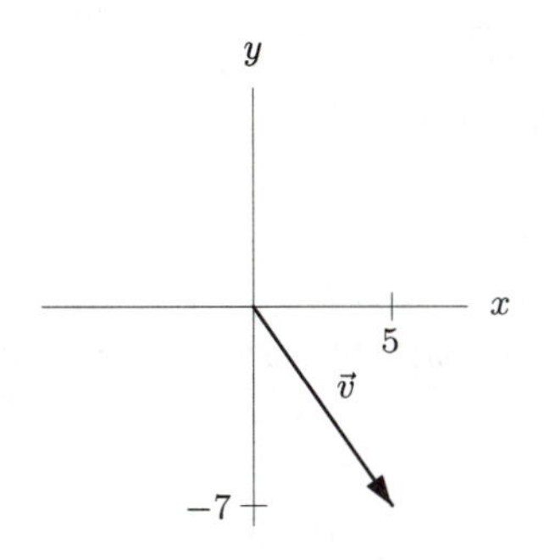

Figure 13.1

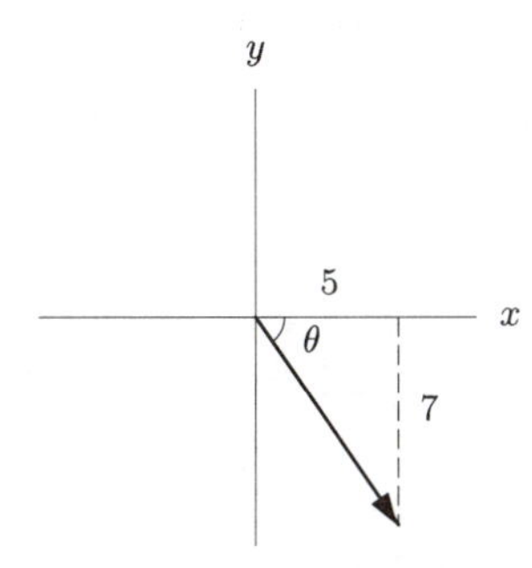

Figure 13.2

Problems

21. **(a)** True, since vectors $\vec{c}$ and $\vec{f}$ point in the same direction and have the same length.
(b) False, since vectors $\vec{a}$ and $\vec{d}$ point in opposite directions. We have $\vec{a} = -\vec{d}$.
(c) False, since $-\vec{b}$ points in the opposite direction to $\vec{b}$, the vectors $-\vec{b}$ and $\vec{a}$ are perpendicular.
(d) True. The vector $\vec{f}$ can be "moved" to point directly up the z-axis.
(e) True. We move in the positive x-direction following vector $\vec{a}$ and then in the positive y-direction following vector $-\vec{b}$. The resulting sum is the vector $\vec{e}$.
(f) False, vector $\vec{d}$ is the negative of the vector $\vec{g} - \vec{c}$. It is true that $\vec{d} = \vec{c} - \vec{g}$.

25.

$$\begin{aligned}\text{Displacement} &= \text{Cat's coordinates} - \text{Bottom of the tree's coordinates}\\ &= (1-2)\vec{i} + (4-4)\vec{j} + (0-0)\vec{k} = -\vec{i}.\end{aligned}$$

29. Since the component of $\vec{v}$ in the $\vec{i}$-direction is 3, we have $\vec{v} = 3\vec{i} + b\vec{j}$ for some b. Since $\|\vec{v}\| = 5$, we have $\sqrt{3^2+b^2} = 5$, so $b = 4$ or $b = -4$. There are two vectors satisfying the properties given: $\vec{v} = 3\vec{i} + 4\vec{j}$ and $\vec{v} = 3\vec{i} - 4\vec{j}$.

33. In Figure 13.3 let O be the origin, points A, B, and C be the vertices of the triangle, point D be the midpoint of $\overline{BC}$, and Q be the point in the line segment $\overline{DA}$ that is $\frac{1}{3}|DA|$ away from D.

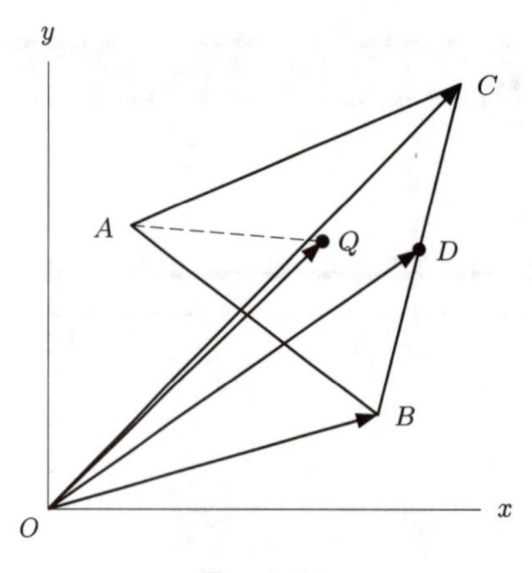

Figure 13.3

From Figure 13.3 we see that

$$\begin{aligned}
\overrightarrow{OQ} = \overrightarrow{OD} + \overrightarrow{DQ} = \overrightarrow{OD} + \frac{1}{3}\overrightarrow{DA} \\
= \overrightarrow{OD} + \frac{1}{3}(\overrightarrow{OA} - \overrightarrow{OD}) \\
= \overrightarrow{OD} + \frac{1}{3}\overrightarrow{OA} - \frac{1}{3}\overrightarrow{OD} \\
= \frac{1}{3}\overrightarrow{OA} + \frac{2}{3}\overrightarrow{OD}.
\end{aligned}$$

Because the diagonals of a parallelogram meet at their midpoint, and $2\overrightarrow{OD}$ is a diagonal of the parallelogram formed by $\overrightarrow{OB}$ and $\overrightarrow{OC}$, we have:

$$\overrightarrow{OD} = \frac{1}{2}(\overrightarrow{OB} + \overrightarrow{OC}),$$

so we can write:

$$\overrightarrow{OQ} = \frac{1}{3}\overrightarrow{OA} + \frac{2}{3}\left(\frac{1}{2}\right)(\overrightarrow{OB} + \overrightarrow{OC}) = \frac{1}{3}(\overrightarrow{OA} + \overrightarrow{OB} + \overrightarrow{OC}).$$

Thus a vector from the origin to a point $\frac{1}{3}$ of the way along median AD from D, the midpoint, is given by $\frac{1}{3}(\overrightarrow{OA} + \overrightarrow{OB} + \overrightarrow{OC})$.

In a similar manner we can show that the vector from the origin to the point $\frac{1}{3}$ of the way along any median from the midpoint of the side it bisects is also $\frac{1}{3}(\overrightarrow{OA} + \overrightarrow{OB} + \overrightarrow{OC})$. See Figure 13.4 and 13.5.

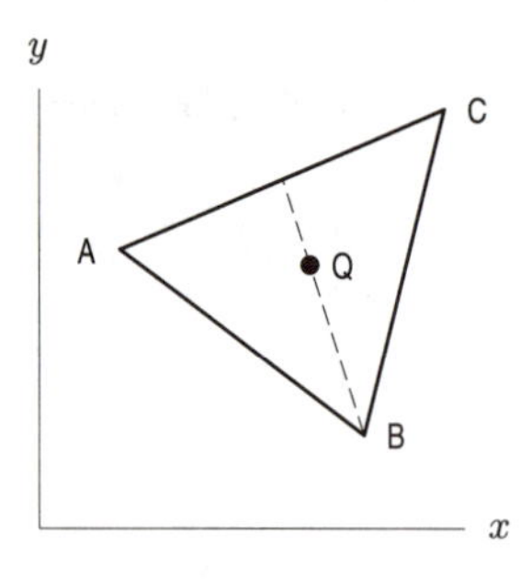

Figure 13.4

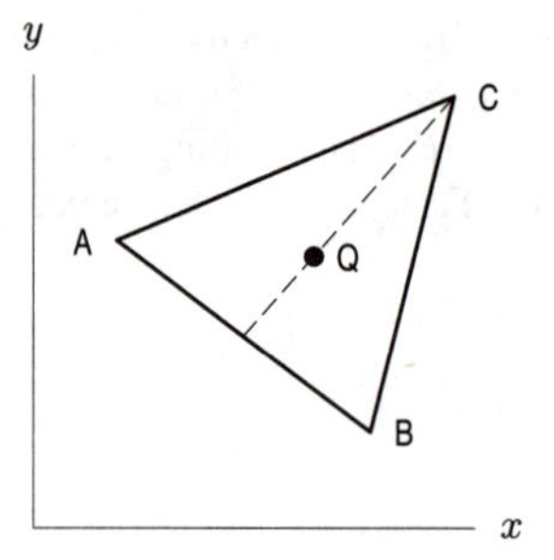

Figure 13.5

Thus the medians of a triangle intersect at a point $\frac{1}{3}$ of the way along each median from the side that each bisects.

Solutions for Section 13.2

Exercises

1. Scalar

5. Writing $\vec{P} = (P_1, P_2, \cdots, P_{50})$ where P_i is the population of the i-th state, shows that $\vec{P}$ can be thought of as a vector with 50 components.

9. We need to calculate the length of each vector.

$$\begin{aligned} \|21\vec{i} + 35\vec{j}\| &= \sqrt{21^2 + 35^2} = \sqrt{1666} \approx 40.8, \\ \|40\vec{i}\| &= \sqrt{40^2} = 40. \end{aligned}$$

So the first car is faster.

Problems

13. **(a)** The velocity vector for the boat is $\vec{b} = 25\vec{i}$ and the velocity vector for the current is

$$\vec{c} = -10\cos(45°)\vec{i} - 10\sin(45°)\vec{j} = -7.07\vec{i} - 7.07\vec{j}.$$

The actual velocity of the boat is

$$\vec{b} + \vec{c} = 17.93\vec{i} - 7.07\vec{j}.$$

(b) $\|\vec{b} + \vec{c}\| = 19.27$ km/hr.

(c) We see in Figure 13.6 that $\tan\theta = \dfrac{7.07}{17.93}$, so $\theta = 21.52°$ south of east.

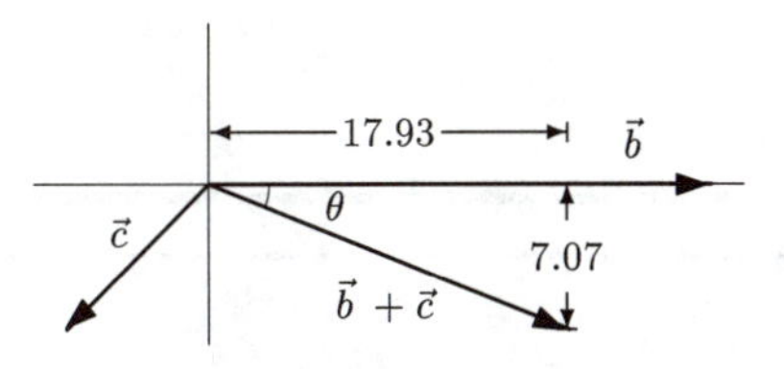

Figure 13.6

17. Let the x-axis point east and the y-axis point north. Since the wind is blowing from the northeast at a speed of 50 km/hr, the velocity of the wind is

$$\vec{w} = -50\cos 45°\vec{i} - 50\sin 45°\vec{j} \approx -35.4\vec{i} - 35.4\vec{j}.$$

Let $\vec{a}$ be the velocity of the airplane, relative to the air, and let ϕ be the angle from the x-axis to $\vec{a}$; since $\|\vec{a}\| =$ 600 km/hr, we have $\vec{a} = 600\cos\phi\vec{i} + 600\sin\phi\vec{j}$. (See Figure 13.7.)

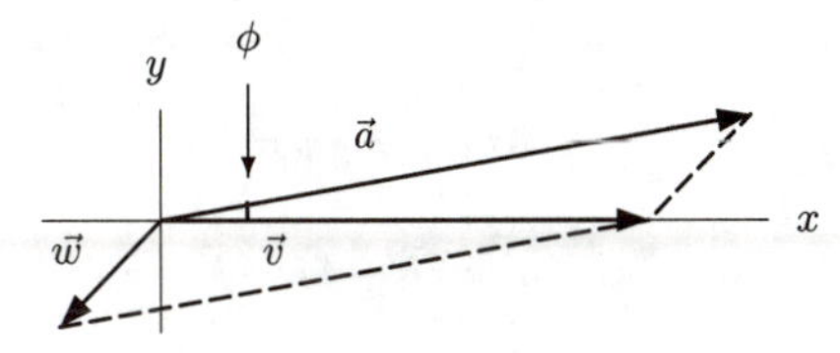

Figure 13.7

Now the resultant velocity, $\vec{v}$, is given by

$$\begin{aligned} \vec{v} = \vec{a} + \vec{w} &= (600\cos\phi\vec{i} + 600\sin\phi\vec{j}) + (-35.4\vec{i} - 35.4\vec{j}) \\ &= (600\cos\phi - 35.4)\vec{i} + (600\sin\phi - 35.4)\vec{j}. \end{aligned}$$

Since the airplane is to fly due east, i.e., in the x direction, then the y-component of the velocity must be 0, so we must have

$$600\sin\phi - 35.4 = 0$$
$$\sin\phi = \frac{35.4}{600}.$$

Thus $\phi = \arcsin(35.4/600) \approx 3.4^\circ$.

21. We want the total force on the object to be zero. We must choose the third force $\vec{F}_3$ so that $\vec{F}_1 + \vec{F}_2 + \vec{F}_3 = 0$. Since $\vec{F}_1 + \vec{F}_2 = 11\vec{i} - 4\vec{j}$, we need $\vec{F}_3 = -11\vec{i} + 4\vec{j}$.

25.

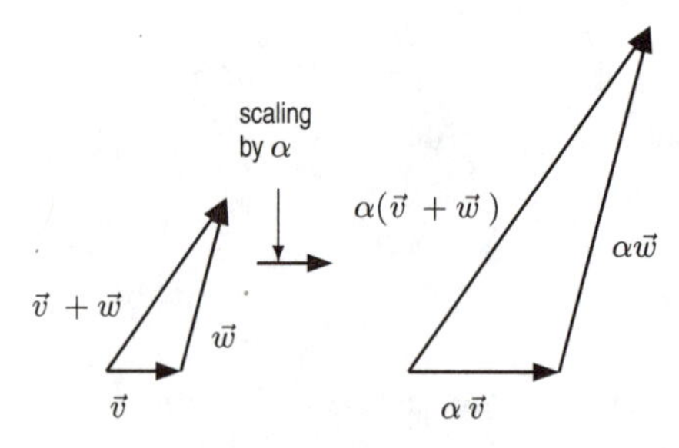

Figure 13.8

The effect of scaling the left-hand picture in Figure 13.8 is to stretch each vector by a factor of α (shown with $\alpha > 1$). Since, after scaling up, the three vectors $\alpha\,\vec{v}$, $\alpha\vec{w}$, and $\alpha(\vec{v} + \vec{w})$ form a similar triangle, we know that $\alpha(\vec{v} + \vec{w})$ is the sum of the other two: that is

$$\alpha(\vec{v} + \vec{w}) = \alpha\,\vec{v} + \alpha\vec{w}.$$

29. According to the definition of scalar multiplication, $1 \cdot \vec{v}$ has the same direction and magnitude as $\vec{v}$, so it is the same as $\vec{v}$.

Solutions for Section 13.3

Exercises

1. $\vec{c} \cdot \vec{y} = (\vec{i} + 6\vec{j}) \cdot (4\vec{i} - 7\vec{j}) = (1)(4) + (6)(-7) = 4 - 42 = -38$.

5. Since $\vec{a} \cdot \vec{y}$ and $\vec{c} \cdot \vec{z}$ are both scalars, the answer to this equation is the product of two numbers and therefore a number. We have

$$\vec{a} \cdot \vec{y} = (2\vec{j} + \vec{k}) \cdot (4\vec{i} - 7\vec{j}) = 0(4) + 2(-7) + 1(0) = -14$$
$$\vec{c} \cdot \vec{z} = (\vec{i} + 6\vec{j}) \cdot (\vec{i} - 3\vec{j} - \vec{k}) = 1(1) + 6(-3) + 0(-1) = -17$$

Thus,

$$(\vec{a} \cdot \vec{y})(\vec{c} \cdot \vec{z}) = 238$$

9. Writing the equation in the form

$$3x + 4y - z = 7$$

shows that a normal vector is

$$\vec{n} = 3\vec{i} + 4\vec{j} - \vec{k}$$

Problems

13. Since $3\vec{i} + \sqrt{3}\vec{j} = \sqrt{3}(\sqrt{3}\vec{i} + \vec{j})$, we know that $3\vec{i} + \sqrt{3}\vec{j}$ and $\sqrt{3}\vec{i} + \vec{j}$ are scalar multiples of one another, and therefore parallel.

Since $(\sqrt{3}\vec{i} + \vec{j}) \cdot (\vec{i} - \sqrt{3}\vec{j}) = \sqrt{3} - \sqrt{3} = 0$, we know that $\sqrt{3}\vec{i} + \vec{j}$ and $\vec{i} - \sqrt{3}\vec{j}$ are perpendicular.
Since $3\vec{i} + \sqrt{3}\vec{j}$ and $\sqrt{3}\vec{i} + \vec{j}$ are parallel, $3\vec{i} + \sqrt{3}\vec{j}$ and $\vec{i} - \sqrt{3}\vec{j}$ are perpendicular, too.

17. Since the plane is normal to the vector $5\vec{i} + \vec{j} - 2\vec{k}$ and passes through the point $(0, 1, -1)$, an equation for the plane is

$$\begin{aligned} 5x + y - 2z &= 5 \cdot 0 + 1 \cdot 1 + (-2) \cdot (-1) = 3 \\ 5x + y - 2z &= 3. \end{aligned}$$

21. **(a)** The plane can be written as $5x - 2y - z + 7 = 0$, so the vector $5\vec{i} - 2\vec{j} - \vec{k}$ is normal to the plane. The vector $\lambda\vec{i} + \vec{j} + 0.5\vec{k}$ is parallel to $5\vec{i} - 2\vec{j} - \vec{k}$ if one is a scalar multiple of the other. This occurs if the coeffients are in proportion:

$$\frac{\lambda}{5} = \frac{1}{-2} = \frac{0.5}{-1}.$$

Solving gives $\lambda = -2.5$.

(b) Substituting $x = a + 1$, $y = a$, $z = a - 1$ into the equation of the plane gives

$$\begin{aligned} a - 1 &= 5(a + 1) - 2a + 7 \\ a - 1 &= 5a + 5 - 2a + 7 \\ -13 &= 2a \\ a &= -6.5. \end{aligned}$$

25. Let

$$\vec{a} = \vec{a}_{\text{parallel}} + \vec{a}_{\text{perp}}$$

where $\vec{a}_{\text{parallel}}$ is parallel to $\vec{d}$, and $\vec{a}_{\text{perp}}$ is perpendicular to $\vec{d}$. Then $\vec{a}_{\text{parallel}}$ is the projection of $\vec{a}$ in the direction of $\vec{d}$:

$$\begin{aligned} \vec{a}_{\text{parallel}} &= \left(\vec{a} \cdot \frac{\vec{d}}{\|\vec{d}\|}\right) \frac{\vec{d}}{\|\vec{d}\|} \\ &= \left((3\vec{i} + 2\vec{j} - 6\vec{k}) \cdot \frac{(2\vec{i} - 4\vec{j} + \vec{k})}{\sqrt{2^2 + 4^2 + 1^2}}\right) \frac{(2\vec{i} - 4\vec{j} + \vec{k})}{\sqrt{2^2 + 4^2 + 1^2}} \\ &= -\frac{8}{21}(2\vec{i} - 4\vec{j} + \vec{k}) \\ &= -\frac{8}{21}\vec{d} \end{aligned}$$

Since we now know $\vec{a}$ and $\vec{a}_{\text{parallel}}$, we can solve for $\vec{a}_{\text{perp}}$:

$$\begin{aligned} \vec{a}_{\text{perp}} &= \vec{a} - \vec{a}_{\text{parallel}} \\ &= (3\vec{i} + 2\vec{j} - 6\vec{k}) - \left(-\frac{8}{21}\right)(2\vec{i} - 4\vec{j} + \vec{k}) \\ &= \frac{79}{21}\vec{i} + \frac{10}{21}\vec{j} - \frac{118}{21}\vec{k}. \end{aligned}$$

Thus we can now write $\vec{a}$ as the sum of two vectors, one parallel to $\vec{d}$, the other perpendicular to $\vec{d}$:

$$\vec{a} = -\frac{8}{21}\vec{d} + \left(\frac{79}{21}\vec{i} + \frac{10}{21}\vec{j} - \frac{118}{21}\vec{k}\right)$$

29. We have

$$\begin{aligned} \vec{p} \cdot \vec{q} &= (1.00)(43) + (3.50)(57) + (4.00)(12) + (2.75)(78) + (5.00)(20) + (3.00)(35) \\ &= 710 \text{ dollars}. \end{aligned}$$

The vendor took in \$710 in from sales. The quantity $\vec{p} \cdot \vec{q}$ represents the total revenue earned.

33. **(a)** The geometric definition of the dot product says that

$$\vec{n} \cdot \overrightarrow{P_0P} = \|\vec{n}\|\|\overrightarrow{P_0P}\| \cos\theta,$$

where θ is the angle between $\vec{n}$ and $\overrightarrow{P_0P}$ with $0 \leq \theta \leq \pi$. To say that the dot product $\vec{n} \cdot \overrightarrow{P_0P}$ is positive means that the angle between $\vec{n}$ and $\overrightarrow{P_0P}$ is between 0 and $\pi/2$, and strictly less than $\pi/2$. Hence $\vec{n}$ and $\overrightarrow{P_0P}$ are both pointing to the same side of the plane. Thus, all the points satisfying $\vec{n} \cdot \overrightarrow{P_0P} > 0$ are on the same side of the plane, the side which $\vec{n}$ points to. To say that the dot product is negative is to say that $\pi/2 < \theta \leq \pi$, and this means that $\overrightarrow{P_0P}$ and $\vec{n}$ are pointing to opposite sides of the plane. Thus, all points satisfying $\vec{n} \cdot \overrightarrow{P_0P} < 0$ are on the side of the plane opposite to $\vec{n}$.

(b) Suppose the normal vector is $\vec{n} = a\vec{i} + b\vec{j} + c\vec{k}$, let $P_0 = (x_0, y_0, z_0)$ be a point in the plane and let $P = (x, y, z)$ be a variable point. Then $\overrightarrow{P_0P} = (x - x_0)\vec{i} + (y - y_0)\vec{j} + (z - z_0)\vec{k}$. Then $\vec{n} \cdot \overrightarrow{P_0P} > 0$ means

$$a(x - x_0) + b(y - y_0) + c(z - z_0) > 0$$

and $\vec{n} \cdot \overrightarrow{P_0P} < 0$ means

$$a(x - x_0) + b(y - y_0) + c(z - z_0) < 0$$

If the equation of the plane is written $ax + by + cz = d$ (with $d = ax_0 + by_0 + cz_0$) then the inequalities become

$$ax + by + cz > d \quad \text{and} \quad ax + by + cz < d.$$

(c) We test each of the points $P = (-1, -1, 1)$, $Q = (-1, -1, -1)$ and $R = (1, 1, 1)$, using the coordinate version of the inequalites in part (b):

$$\begin{aligned} P: &\quad 2 \cdot (-1) - 3 \cdot (-1) + 4 \cdot 1 = 5 > 4 \\ Q: &\quad 2 \cdot (-1) - 3 \cdot (-1) + 4 \cdot (-1) = -3 < 4 \\ R: &\quad 2 \cdot 1 - 3 \cdot 1 + 4 \cdot 1 = 3 < 4 \end{aligned}$$

Therefore Q and R are on the same side of the plane as each other; P is on the other side.

37. Since $\vec{u} \cdot \vec{w} = \vec{v} \cdot \vec{w}$, $(\vec{u} - \vec{v}) \cdot \vec{w} = 0$. This equality holds for any $\vec{w}$, so we can take $\vec{w} = \vec{u} - \vec{v}$. This gives

$$\|\vec{u} - \vec{v}\|^2 = (\vec{u} - \vec{v}) \cdot (\vec{u} - \vec{v}) = 0,$$

that is,

$$\|\vec{u} - \vec{v}\| = 0.$$

This implies $\vec{u} - \vec{v} = 0$, that is, $\vec{u} = \vec{v}$.

41. We substitute $\vec{u} = u_1\vec{i} + u_2\vec{j} + u_3\vec{k}$ and by the result of Problem 38, we expand as follows:

$$\begin{aligned} (\vec{u} \cdot \vec{v})_{\text{geom}} &= (u_1\vec{i} + u_2\vec{j} + u_3\vec{k}) \cdot \vec{v} \\ &= (u_1\vec{i}) \cdot \vec{v} + (u_2\vec{j}) \cdot \vec{v} + (u_3\vec{k}) \cdot \vec{v} \end{aligned}$$

where all the dot products are defined geometrically By the result of Problem 39 we can write

$$(\vec{u} \cdot \vec{v})_{\text{geom}} = u_1(\vec{i} \cdot \vec{v})_{\text{geom}} + u_2(\vec{j} \cdot \vec{v})_{\text{geom}} + u_3(\vec{k} \cdot \vec{v})_{\text{geom}}.$$

Now substitute $\vec{v} = v_1\vec{i} + v_2\vec{j} + v_3\vec{k}$ and expand, again using Problem 38 and the geometric definition of the dot product:

$$\begin{aligned} (\vec{u} \cdot \vec{v})_{\text{geom}} &= u_1\left(\vec{i} \cdot (v_1\vec{i} + v_2\vec{j} + v_3\vec{k})\right)_{\text{geom}} \\ &\quad + u_2\left(\vec{j} \cdot (v_1\vec{i} + v_2\vec{j} + v_3\vec{k})\right)_{\text{geom}} \\ &\quad + u_3\left(\vec{k} \cdot (v_1\vec{i} + v_2\vec{j} + v_3\vec{k})\right)_{\text{geom}} \\ &= u_1v_1(\vec{i} \cdot \vec{i})_{\text{geom}} + u_1v_2(\vec{i} \cdot \vec{j})_{\text{geom}} + u_1v_3(\vec{i} \cdot \vec{k})_{\text{geom}} \\ &\quad + u_2v_1(\vec{i} \cdot \vec{i})_{\text{geom}} + u_2v_2(\vec{i} \cdot \vec{j})_{\text{geom}} + u_2v_3(\vec{i} \cdot \vec{k})_{\text{geom}} \\ &\quad + u_3v_1(\vec{i} \cdot \vec{i})_{\text{geom}} + u_3v_2(\vec{i} \cdot \vec{j})_{\text{geom}} + u_3v_3(\vec{i} \cdot \vec{k})_{\text{geom}} \end{aligned}$$

The geometric definition of the dot product shows that

$$\vec{i}\cdot\vec{i} = \|\vec{i}\|\,\|\vec{i}\|\cos 0 = 1$$
$$\vec{i}\cdot\vec{j} = \|\vec{i}\|\,\|\vec{j}\|\cos\frac{\pi}{2} = 0.$$

Similarly $\vec{j}\cdot\vec{j} = \vec{k}\cdot\vec{k} = 1$ and $\vec{i}\cdot\vec{k} = \vec{j}\cdot\vec{k} = 0$. Thus, the expression for $(\vec{u}\cdot\vec{v})_{\text{geom}}$ becomes

$$\begin{aligned}(\vec{u}\cdot\vec{v})_{\text{geom}} &= u_1v_1(1) + u_1v_2(0) + u_1v_3(0)\\ &\quad+u_2v_1(0) + u_2v_2(1) + u_2v_3(0)\\ &\quad+u_3v_1(0) + u_3v_2(0) + u_3v_3(1)\\ &= u_1v_1 + u_2v_2 + u_3v_3.\end{aligned}$$

Solutions for Section 13.4

Exercises

1. $\vec{v}\times\vec{w} = \vec{k}\times\vec{j} = -\vec{i}$ (remember $\vec{i},\vec{j},\vec{k}$ are unit vectors along the axes, and you must use the right hand rule.)

5. $\vec{v} = 2\vec{i} - 3\vec{j} + \vec{k}$, and $\vec{w} = \vec{i} + 2\vec{j} - \vec{k}$

$$\vec{v}\times\vec{w} = \begin{vmatrix}\vec{i} & \vec{j} & \vec{k}\\ 2 & -3 & 1\\ 1 & 2 & -1\end{vmatrix} = \vec{i} + 3\vec{j} + 7\vec{k}$$

9.

$$\begin{aligned}(\vec{i}+\vec{j})\times(\vec{i}\times\vec{j}) &= (\vec{i}+\vec{j})\times\vec{k}\\ &= (\vec{i}\times\vec{k}) + (\vec{j}\times\vec{k})\\ &= -\vec{j} + \vec{i} = \vec{i} - \vec{j}.\end{aligned}$$

Problems

13. Since $\vec{v}\times\vec{w}$ is perpendicular to both $\vec{v}$ and $\vec{w}$, we can conclude that $\vec{v}\times\vec{w}$ is parallel to the z-axis.

17. **(a)** If we let $\overrightarrow{PQ}$ in Figure 13.9 be the vector from point P to point Q and $\overrightarrow{PR}$ be the vector from P to R, then

$$\overrightarrow{PQ} = -\vec{i} + 2\vec{k}$$
$$\overrightarrow{PR} = 2\vec{i} - \vec{k},$$

then the area of the parallelogram determined by $\overrightarrow{PQ}$ and $\overrightarrow{PR}$ is:

$$\begin{matrix}\text{Area of}\\ \text{parallelogram}\end{matrix} = \|\overrightarrow{PQ}\times\overrightarrow{PR}\| = \left\|\begin{vmatrix}\vec{i} & \vec{j} & \vec{k}\\ -1 & 0 & 2\\ 2 & 0 & -1\end{vmatrix}\right\| = \|3\vec{j}\| = 3.$$

Thus, the area of the triangle PQR is

$$\begin{pmatrix}\text{Area of}\\ \text{triangle}\end{pmatrix} = \frac{1}{2}\begin{pmatrix}\text{Area of}\\ \text{parallelogram}\end{pmatrix} = \frac{3}{2} = 1.5.$$

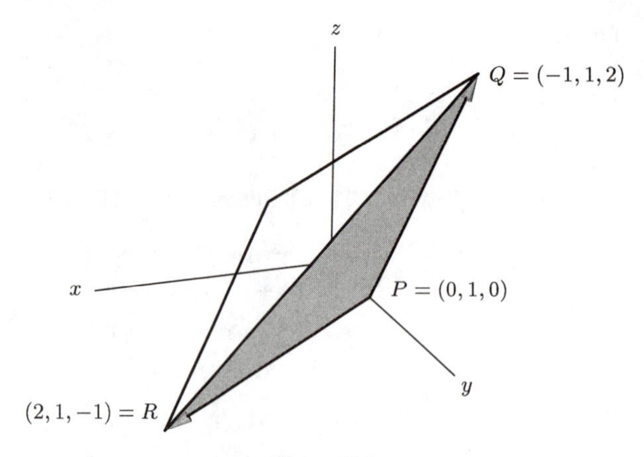

Figure 13.9

(b) Since $\vec{n} = \overrightarrow{PQ} \times \overrightarrow{PR}$ is perpendicular to the plane PQR, and from above, we have $\vec{n} = 3\vec{j}$, the equation of the plane has the form $3y = C$. At the point $(0, 1, 0)$ we get $3 = C$, therefore $3y = 3$, i.e., $y = 1$.

21. We use the same normal vector $\vec{n} = 4\vec{i} + 26\vec{j} + 14\vec{k}$ and the point $(4, 5, 6)$ to get $4(x-4)+26(y-5)+14(z-6) = 0$.

25. If $\lambda = 0$, then all three cross products are $\vec{0}$, since the cross product of the zero vector with any other vector is always 0.

If $\lambda > 0$, then $\lambda\vec{v}$ and $\vec{v}$ are in the same direction and $\vec{w}$ and $\lambda\vec{w}$ are in the same direction. Therefore the unit normal vector $\vec{n}$ is the same in all three cases. In addition, the angles between $\lambda\vec{v}$ and $\vec{w}$, and between $\vec{v}$ and $\vec{w}$, and between $\vec{v}$ and $\lambda\vec{w}$ are all θ. Thus,

$$\begin{aligned}
(\lambda\vec{v}) \times \vec{w} &= \|\lambda\vec{v}\|\|\vec{w}\| \sin\theta\,\vec{n} \\
&= \lambda\|\vec{v}\|\|\vec{w}\| \sin\theta\,\vec{n} \\
&= \lambda(\vec{v} \times \vec{w}) \\
&= \|\vec{v}\|\|\lambda\vec{w}\| \sin\theta\,\vec{n} \\
&= \vec{v} \times (\lambda\vec{w})
\end{aligned}$$

If $\lambda < 0$, then $\lambda\vec{v}$ and $\vec{v}$ are in opposite directions, as are $\vec{w}$ and $\lambda\vec{w}$ in opposite directions. Therefore if $\vec{n}$ is the normal vector in the definition of $\vec{v} \times \vec{w}$, then the right-hand rule gives $-\vec{n}$ for $(\lambda\vec{v}) \times \vec{w}$ and $\vec{v} \times (\lambda\vec{w})$. In addition, if the angle between $\vec{v}$ and $\vec{w}$ is θ, then the angle between $\lambda\vec{v}$ and $\vec{w}$ and between $\vec{v}$ and $\lambda\vec{w}$ is $(\pi - \theta)$. Since if $\lambda < 0$, we have $|\lambda| = -\lambda$, so

$$\begin{aligned}
(\lambda\vec{v}) \times \vec{w} &= \|\lambda\vec{v}\|\|\vec{w}\| \sin(\pi-\theta)(-\vec{n}) \\
&= |\lambda|\,\|\vec{v}\|\|\vec{w}\| \sin(\pi-\theta)(-\vec{n}) \\
&= -\lambda\|\vec{v}\|\|\vec{w}\| \sin\theta(-\vec{n}) \\
&= \lambda\|\vec{v}\|\|\vec{w}\| \sin\theta\,\vec{n} \\
&= \lambda(\vec{v} \times \vec{w}).
\end{aligned}$$

Similarly,

$$\begin{aligned}
\vec{v} \times (\lambda\vec{w}) &= \|\vec{v}\|\|\lambda\vec{w}\| \sin(\pi-\theta)(-\vec{n}) \\
&= -\lambda\|\vec{v}\|\|\vec{w}\| \sin\theta(-\vec{n}) \\
&= \lambda(\vec{v} \times \vec{w}).
\end{aligned}$$

29. Problem 26 tells us that $(\vec{u} \times \vec{v}) \cdot \vec{w} = \vec{u} \cdot (\vec{v} \times \vec{w})$. Using this result on the triple product of $(\vec{a} + \vec{b}) \times \vec{c}$ with any vector $\vec{d}$ together with the fact that the dot product distributes over addition gives us:

$$\begin{aligned}
[(\vec{a} + \vec{b}) \times \vec{c}] \cdot \vec{d} &= (\vec{a} + \vec{b}) \cdot (\vec{c} \times \vec{d}) \\
&= \vec{a} \cdot (\vec{c} \times \vec{d}) + \vec{b} \cdot (\vec{c} \times \vec{d}) && \text{(dot product is distributive)} \\
&= (\vec{a} \times \vec{c}) \cdot \vec{d} + (\vec{b} \times \vec{c}) \cdot \vec{d} && \text{(using Problem 26 again)} \\
&= [(\vec{a} \times \vec{c}) + (\vec{b} \times \vec{c})] \cdot \vec{d}. && \text{(dot product is distributive)}
\end{aligned}$$

So, since $[(\vec{a}+\vec{b})\times\vec{c}]\cdot\vec{d} = [(\vec{a}\times\vec{c})+(\vec{b}\times\vec{c})]\cdot\vec{d}$, then

$$[(\vec{a}+\vec{b})\times\vec{c}]\cdot\vec{d} - [(\vec{a}\times\vec{c})+(\vec{b}\times\vec{c})]\cdot\vec{d} = 0,$$

Since the dot product is distributive, we have

$$[((\vec{a}+\vec{b})\times\vec{c}) - (\vec{a}\times\vec{c}) - (\vec{b}\times\vec{c})]\cdot\vec{d} = 0.$$

Since this equation is true for all vectors $\vec{d}$, by letting

$$\vec{d} = ((\vec{a}+\vec{b})\times\vec{c}) - (\vec{a}\times\vec{c}) - (\vec{b}\times\vec{c}),$$

we get

$$\|(\vec{a}+\vec{b})\times\vec{c} - \vec{a}\times\vec{c} - \vec{b}\times\vec{c}\|^2 = 0$$

and hence

$$(\vec{a}+\vec{b})\times\vec{c} - (\vec{a}\times\vec{c}) - (\vec{b}\times\vec{c}) = \vec{0}.$$

Thus

$$(\vec{a}+\vec{b})\times\vec{c} = (\vec{a}\times\vec{c}) + (\vec{b}\times\vec{c}).$$

Solutions for Chapter 13 Review

Exercises

1. $\vec{v}+2\vec{w} = 2\vec{i}+3\vec{j}-\vec{k}+2(\vec{i}-\vec{j}+2\vec{k}) = 4\vec{i}+\vec{j}+3\vec{k}$.

5. Since $\vec{v}\cdot\vec{w} = 2\cdot 1+3(-1)+(-1)2 = -3$, we have $(\vec{v}\cdot\vec{w})\vec{v} = -6\vec{i}-9\vec{j}+3\vec{k}$.

9. **(a)** We have $\vec{v}\cdot\vec{w} = 3\cdot 4+2\cdot(-3)+(-2)\cdot 1 = 4$.

(b) We have $\vec{v}\times\vec{w} = -4\vec{i}-11\vec{j}-17\vec{k}$.

(c) A vector of length 5 parallel to $\vec{v}$ is

$$\frac{5}{\|\vec{v}\|}\vec{v} = \frac{5}{\sqrt{17}}(3\vec{i}+2\vec{j}-2\vec{k}) = 3.64\vec{i}+2.43\vec{j}-2.43\vec{k}.$$

(d) The angle between vectors $\vec{v}$ and $\vec{w}$ is found using

$$\cos\theta = \frac{\vec{v}\cdot\vec{w}}{\|\vec{v}\|\|\vec{w}\|} = \frac{4}{\sqrt{17}\sqrt{26}} = 0.190,$$

so $\theta = 79.0^\circ$.

(e) The component of vector $\vec{v}$ in the direction of vector $\vec{w}$ is

$$\frac{\vec{v}\cdot\vec{w}}{\|\vec{w}\|} = \frac{4}{\sqrt{26}} = 0.784.$$

(f) The answer is any vector $\vec{a}$ such that $\vec{a}\cdot\vec{v} = 0$. One possible answer is $2\vec{i}-2\vec{j}+\vec{k}$.

(g) A vector perpendicular to both is the cross product:

$$\vec{v}\times\vec{w} = -4\vec{i}-11\vec{j}-17\vec{k}.$$

13. The vector $\vec{w}$ we want is shown in Figure 13.10, where the given vector is $\vec{v} = 4\vec{i}+3\vec{j}$. The vectors $\vec{v}$ and $\vec{w}$ are the same length and the two angles marked α are equal, so the two right triangles shown are congruent. Thus

$$a = -3 \quad \text{and} \quad b = 4.$$

Therefore

$$\vec{w} = -3\vec{i}+4\vec{j}.$$

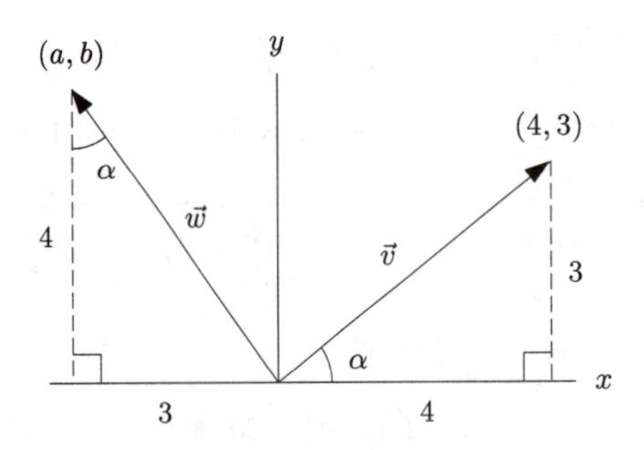

Figure 13.10

Problems

17. **(a)** On the x-axis, $y = z = 0$, so $5x = 21$, giving $x = \frac{21}{5}$. So the only such point is $(\frac{21}{5}, 0, 0)$.
(b) Other points are $(0, -21, 0)$, and $(0, 0, 3)$. There are many other possible answers.
(c) $\vec{n} = 5\vec{i} - \vec{j} + 7\vec{k}$. It is the normal vector.
(d) The vector between two points in the plane is parallel to the plane. Using the points from part (b), the vector $3\vec{k} - (-21\vec{j}) = 21\vec{j} + 3\vec{k}$ is parallel to the plane.

21. The speed is a scalar which equals 30 times the circumference of the circle per minute. So it is a constant. The velocity is a vector. Since the direction of the motion changes all the time, the velocity is not constant. This implies that the acceleration is nonzero.

25. Let the x-axis point east and the y-axis point north. Denote the forces exerted by Charlie, Sam and Alice by $\vec{F}_C, \vec{F}_S$ and $\vec{F}_A$ (see Figure 13.11).

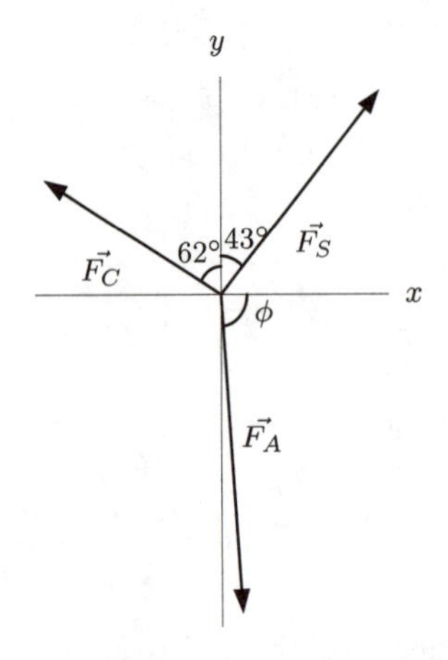

Figure 13.11

Since $\|\vec{F}_C\| = 175$ newtons and the angle θ from the x-axis to $\vec{F}_C$ is $90° + 62° = 152°$, we have

$$\vec{F}_C = 175\cos 152°\vec{i} + 175\sin 152°\vec{j} \approx -154.52\vec{i} + 82.16\vec{j}.$$

Similarly,

$$\vec{F}_S = 200\cos 47°\vec{i} + 200\sin 47°\vec{j} \approx 136.4\vec{i} + 146.27\vec{j}.$$

Now Alice is to counterbalance Sam and Charlie, so the resultant force of the three forces $\vec{F}_C, \vec{F}_S$ and $\vec{F}_A$ must be 0, that is,

$$\vec{F}_C + \vec{F}_S + \vec{F}_A = 0.$$

Thus, we have

$$\begin{aligned}\vec{F}_A &= -\vec{F}_C - \vec{F}_S \\ &\approx -(-154.52\vec{i} + 82.16\vec{j}) - (136.4\vec{i} + 146.27\vec{j}) \\ &= 18.12\vec{i} - 228.43\vec{j}\end{aligned}$$

and, $||\vec{F}_A|| = \sqrt{18.12^2 + (-228.43)^2} \approx 229.15$ newtons.

If ϕ is the angle from the x-axis to $\vec{F}_A$, then

$$\phi = \arctan\frac{-228.43}{18.12} \approx -85.5^\circ.$$

29. The displacement from $(1,1,1)$ to $(1,4,5)$ is

$$\vec{r_1} = (1-1)\vec{i} + (4-1)\vec{j} + (5-1)\vec{k} = 3\vec{j} + 4\vec{k}.$$

The displacement from $(-3,-2,0)$ to $(1,4,5)$ is

$$\vec{r_2} = (1+3)\vec{i} + (4+2)\vec{j} + (5-0)\vec{k} = 4\vec{i} + 6\vec{j} + 5\vec{k}.$$

A normal vector is

$$\vec{n} = \vec{r_1} \times \vec{r_2} = \begin{vmatrix} \vec{i} & \vec{j} & \vec{k} \\ 0 & 3 & 4 \\ 4 & 6 & 5 \end{vmatrix} = (15-24)\vec{i} - (-16)\vec{j} + (-12)\vec{k} = -9\vec{i} + 16\vec{j} - 12\vec{k}.$$

The equation of the plane is

$$-9x + 16y - 12z = -9\cdot 1 + 16\cdot 1 - 12\cdot 1 = -5$$
$$9x - 16y + 12z = 5.$$

We pick a point A on the plane, $A = (\frac{5}{9}, 0, 0)$ and let $P = (0,0,0)$. (See Figure 13.12.) Then $\vec{PA} = (5/9)\vec{i}$.

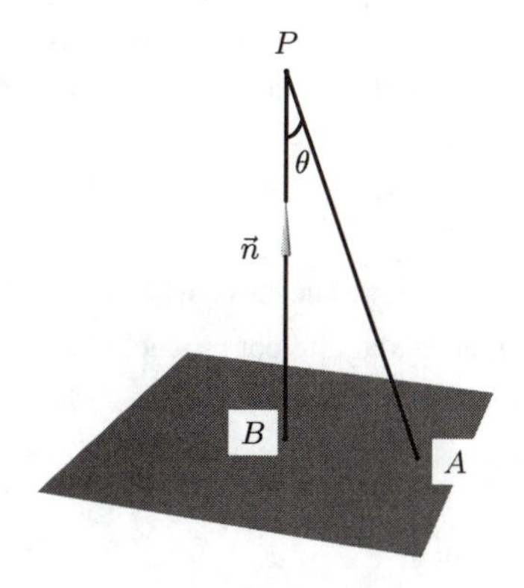

Figure 13.12

So the distance d from the point P to the plane is

$$\begin{aligned} d &= ||\vec{PB}|| = ||\vec{PA}||\cos\theta \\ &= \frac{\vec{PA}\cdot\vec{n}}{||\vec{n}||} \quad \text{since } \vec{PA}\cdot\vec{n} = ||\vec{PA}||\,||\vec{n}||\cos\theta) \\ &= \left|\frac{(\frac{5}{9}\vec{i})\cdot(-9\vec{i} + 16\vec{j} - 12\vec{k})}{\sqrt{9^2+16^2+12^2}}\right| \\ &= \frac{5}{\sqrt{481}} = 0.23. \end{aligned}$$

CAS Challenge Problems

33. $(\vec{a} \times \vec{b}) \cdot \vec{c} = 0$, $(\vec{a} \times \vec{b}) \times (\vec{a} \times \vec{c}) = \vec{0}$

Since $\vec{c}$ is the sum of a scalar multiple of $\vec{a}$ and a scalar multiple of $\vec{b}$, it lies in the plane containing $\vec{a}$ and $\vec{b}$. On the other hand, $\vec{a} \times \vec{b}$ is perpendicular to this plane, so $\vec{a} \times \vec{b}$ is perpendicular to $\vec{c}$. Therefore, $(\vec{a} \times \vec{b}) \cdot \vec{c} = 0$. Also, $\vec{a} \times \vec{c}$ is also perpendicular to the plane, thus parallel to $\vec{a} \times \vec{b}$, and thus $(\vec{a} \times \vec{b}) \times (\vec{a} \times \vec{c}) = \vec{0}$.

37. **(a)** $\overrightarrow{PQ} \times \overrightarrow{PR}$ is perpendicular to the plane containing P, Q, R, and therefore parallel to the normal vector $a\vec{i} + b\vec{j} + c\vec{k}$.

(b)

$$\overrightarrow{PQ} \times \overrightarrow{PR} = (tv - sw - ty + wy + sz - vz)\vec{i} + (-tu + rw + tx - wx - rz + uz)\vec{j} + (su - rv - sx + vx + ry - uy)\vec{k}$$

(c) After substituting $z = (d - ax - by)/c$, $w = (d - au - bv)/c$, $t = (d - ar - bs)/c$ into the result of part (a), and simplifying the expression, we obtain:

$$\begin{aligned}\overrightarrow{PQ} \times \overrightarrow{PR} &= \frac{a(s(u-x) + vx - uy + r(-v+y))}{c}\vec{i} + \\ &\frac{b(s(u-x) + vx - uy + r(-v+y))}{c}\vec{j} + (s(u-x) + vx - uy + r(-v+y))\vec{k} \\ &= \frac{(s(u-x) + vx - uy + r(-v+y))}{c}(a\vec{i} + b\vec{j} + c\vec{k}).\end{aligned}$$

Thus $\overrightarrow{PQ} \times \overrightarrow{PR}$ is a scalar multiple of $a\vec{i} + b\vec{j} + c\vec{k}$, and hence parallel to it.

CHECK YOUR UNDERSTANDING

1. False. There are exactly two unit vectors: one in the same direction as $\vec{v}$ and the other in the opposite direction. Explicitly, the unit vectors parallel to $\vec{v}$ are $\pm\frac{1}{\|\vec{v}\|}\vec{v}$.

5. False. If $\vec{v}$ and $\vec{w}$ are not parallel, the three vectors $\vec{v}$, $\vec{w}$ and $\vec{v} - \vec{w}$ can be thought of as three sides of a triangle. (If the tails of $\vec{v}$ and $\vec{w}$ are placed together, then $\vec{v} - \vec{w}$ is a vector from the head of $\vec{w}$ to the head of $\vec{v}$.) The length of one side of a triangle is less than the sum of the lengths of the other two sides. Alternatively, a counterexample is $\vec{v} = \vec{i}$ and $\vec{w} = \vec{j}$. Then $\|\vec{i} - \vec{j}\| = \sqrt{2}$ but $\|\vec{i}\| - \|\vec{j}\| = 0$.

9. False. To find the displacement vector *from* $(1,1,1)$ *to* $(1,2,3)$ we subtract $\vec{i} + \vec{j} + \vec{k}$ from $\vec{i} + 2\vec{j} + 3\vec{k}$ to get $(1-1)\vec{i} + (2-1)\vec{j} + (3-1)\vec{k} = \vec{j} + 2\vec{k}$.

13. True. The cosine of the angle between the vectors is negative when the angle is between $\pi/2$ and π.

17. False. If the vectors are nonzero and perpendicular, the dot product will be zero (e.g. $\vec{i} \cdot \vec{j} = 0$).

21. True. The cross product yields a vector.

25. False. If $\vec{u}$ and $\vec{w}$ are two different vectors both of which are parallel to $\vec{v}$, then $\vec{v} \times \vec{u} = \vec{v} \times \vec{w} = \vec{0}$, but $\vec{u} \neq \vec{w}$. A counterexample is $\vec{v} = \vec{i}$, $\vec{u} = 2\vec{i}$ and $\vec{w} = 3\vec{i}$.

29. True. Any vector $\vec{w}$ that is parallel to $\vec{v}$ will give $\vec{v} \times \vec{w} = \vec{0}$.

CHAPTER FOURTEEN

Solutions for Section 14.1

Exercises

1. If h is small, then

$$f_x(3,2) \approx \frac{f(3+h,2)-f(3,2)}{h}.$$

With $h = 0.01$, we find

$$f_x(3,2) \approx \frac{f(3.01,2)-f(3,2)}{0.01} = \frac{\frac{3.01^2}{(2+1)} - \frac{3^2}{(2+1)}}{0.01} = 2.00333.$$

With $h = 0.0001$, we get

$$f_x(3,2) \approx \frac{f(3.0001,2)-f(3,2)}{0.0001} = \frac{\frac{3.0001^2}{(2+1)} - \frac{3^2}{(2+1)}}{0.0001} = 2.0000333.$$

Since the difference quotient seems to be approaching 2 as h gets smaller, we conclude

$$f_x(3,2) \approx 2.$$

To estimate $f_y(3,2)$, we use

$$f_y(3,2) \approx \frac{f(3,2+h)-f(3,2)}{h}.$$

With $h = 0.01$, we get

$$f_y(3,2) \approx \frac{f(3,2.01)-f(3,2)}{0.01} = \frac{\frac{3^2}{(2.01+1)} - \frac{3^2}{(2+1)}}{0.01} = -0.99668.$$

With $h = 0.0001$, we get

$$f_y(3,2) \approx \frac{f(3,2.0001)-f(3,2)}{0.0001} = \frac{\frac{3^2}{(2.0001+1)} - \frac{3^2}{(2+1)}}{0.0001} = -0.9999667.$$

Thus, it seems that the difference quotient is approaching -1, so we estimate

$$f_y(3,2) \approx -1.$$

5. **(a)** We expect f_p to be negative because if the price of the product increases, the sales usually decrease.
(b) If the price of the product is \$8 per unit and if \$12000 has been spent on advertising, sales increase by approximately 150 units if an additional \$1000 is spent on advertising.

9. For $f_w(10,25)$ we get

$$f_w(10,25) \approx \frac{f(10+h,25)-f(10,25)}{h}.$$

Choosing $h = 5$ and reading values from Table 12.3 on page 567 of the text, we get

$$f_w(10,25) \approx \frac{f(15,25)-f(10,25)}{5} = \frac{2-10}{5} = -1.6$$

This means that when the wind speed is 10 mph and the true temperature is 25°F, as the wind speed increases from 10 mph by 1 mph we feel a 1.6°F drop in temperature. This rate is negative because the temperature you feel drops as the wind speed increases.

Problems

13. **(a)** For points near the point $(0,5,3)$, moving in the positive x direction, the surface is sloping down and the function is decreasing. Thus, $f_x(0,5) < 0$.

(b) Moving in the positive y direction near this point the surface slopes up as the function increases, so $f_y(0,5) > 0$.

17. **(a)** Estimate $\partial P/\partial r$ and $\partial P/\partial L$ by using difference quotients and reading values of P from the graph:

$$\frac{\partial P}{\partial r}(8,4000) \approx \frac{P(16,4000)-P(8,4000)}{16-8} = \frac{100-80}{8} = 2.5,$$

and

$$\frac{\partial P}{\partial L} \approx \frac{P(8,5000)-P(8,4000)}{5000-4000} = \frac{100-80}{1000} = 0.02.$$

$P_r(8,4000) \approx 2.5$ means that at an interest rate of 8% and a loan amount of \$4000 the monthly payment increases by approximately \$2.50 for every one percent increase of the interest rate. $P_L(8,4000) \approx 0.02$ means the monthly payment increases by approximately \$0.02 for every \$1 increase in the loan amount at an 8% rate and a loan amount of \$4000.

(b) Using difference quotients and reading from the graph

$$\frac{\partial P}{\partial r}(8,6000) \approx \frac{P(14,6000)-P(8,6000)}{14-8} = \frac{140-120}{6} = 3.33,$$

and

$$\frac{\partial P}{\partial L}(8,6000) \approx \frac{P(8,7000)-P(8,6000)}{7000-6000} = \frac{140-120}{1000} = 0.02.$$

Again, we see that the monthly payment increases with increases in interest rate and loan amount. The interest rate is $r = 8\%$ as in part (a), but here the loan amount is $L = \$6000$. Since $P_L(8,4000) \approx P_L(8,6000)$, the increase in monthly payment per unit increase in loan amount remains the same as in part a). However, in this case, the effect of the interest rate is different: here the monthly payment increases by approximately \$3.33 for every one percent increase of interest rate at $r = 8\%$ and loan amount of \$6000.

(c)

$$\frac{\partial P}{\partial r}(13,7000) \approx \frac{P(19,7000)-P(13,7000)}{19-13} = \frac{180-160}{6} = 3.33,$$

and

$$\frac{\partial P}{\partial L}(13,7000) \approx \frac{P(13,8000)-P(13,7000)}{8000-7000} = \frac{180-160}{1000} = 0.02.$$

The figures show that the rates of change of the monthly payment with respect to the interest rate and loan amount are roughly the same for $(r,L) = (8,6000)$ and $(r,L) = (13,7000)$.

21.

Table 14.1 *Estimated values of $H(T, w)$ (in calories/meter^3)*

		w (gm/m^3)			
		0.1	0.2	0.3	0.4
T (°C)	10	110	240	330	450
	20	100	180	260	350
	30	70	150	220	300
	40	65	140	200	270

Values of H from the graph are given in Table 14.1. In order to compute $H_w(T, w)$ for $w = 0.3$, it is useful to have the column corresponding to $w = 0.4$. The row corresponding to $T = 40$ is not used in this problem. The partial derivative $H_w(T, w)$ can be approximated by

$$H_w(10, 0.1) \approx \frac{H(10, 0.1 + h) - H(10, 0.1)}{h} \quad \text{for small } h.$$

We choose $h = 0.1$ because we can read off a value for $H(10, 0.2)$ from the graph. If we take $H(10,\ 0.2) = 240$, we get the approximation

$$H_w(10, 0.1) \approx \frac{H(10, 0.2) - H(10, 0.1)}{0.1} = \frac{240 - 110}{0.1} = 1300.$$

In practical terms, we have found that for fog at 10° C containing 0.1 g/m^3 of water, an increase in the water content of the fog will increase the heat requirement for dissipating the fog at the rate given by $H_w(10, 0.1)$. Specifically, a 1 g/m^3 increase in the water content will increase the heat required to dissipate the fog by about 1300 calories per cubic meter of fog. Wetter fog is harder to dissipate. Other values of $H_w(T, w)$ in Table 14.2 are computed using the formula

$$H_w(T, w) \approx \frac{H(T, w + 0.1) - H(T, w)}{0.1},$$

where we have used Table 14.1 to evaluate H.

Table 14.2 *Table of values of $H_w(T, w)$ (in cal/gm)*

		w (gm/m^3)		
		0.1	0.2	0.3
T (°C)	10	1300	900	1200
	20	800	800	900
	30	800	700	800

Solutions for Section 14.2

Exercises

1. $f_x(x, y) = 10xy^3 + 8y^2 - 6x$ and $f_y(x, y) = 15x^2y^2 + 16xy$.

5. $V_r = \frac{2}{3}\pi rh$

9. $g_x(x, y) = \frac{\partial}{\partial x}\ln(ye^{xy}) = (ye^{xy})^{-1}\frac{\partial}{\partial x}(ye^{xy}) = (ye^{xy})^{-1} \cdot y\frac{\partial}{\partial x}(e^{xy}) = (ye^{xy})^{-1} \cdot y \cdot y \cdot e^{xy} = y$

13. $\frac{\partial}{\partial m}\left(\frac{1}{2}mv^2\right) = \frac{1}{2}v^2$

17. $\frac{\partial V}{\partial r} = \frac{8}{3}\pi rh$ and $\frac{\partial V}{\partial h} = \frac{4}{3}\pi r^2$.

21. $\frac{\partial}{\partial T}\left(\frac{2\pi r}{T}\right) = -\frac{2\pi r}{T^2}$

25. $\frac{\partial T}{\partial l} = \frac{2\pi}{\sqrt{g}} \cdot \frac{1}{2}l^{-1/2} = \frac{\pi}{\sqrt{lg}}$.

29. $\frac{\partial}{\partial M}\left(\frac{2\pi r^{3/2}}{\sqrt{GM}}\right) = 2\pi r^{3/2}(-\frac{1}{2})(GM)^{-3/2}(G) = -\pi r^{3/2}\cdot\frac{G}{GM\sqrt{GM}} = -\frac{\pi r^{3/2}}{M\sqrt{GM}}$

33. We regard x as constant and differentiate with respect to y using the product rule:

$$\frac{\partial z}{\partial y} = 2e^{x+2y}\sin y + e^{x+2y}\cos y$$

Substituting $x = 1$, $y = 0.5$ gives

$$\left.\frac{\partial z}{\partial y}\right|_{(1,0.5)} = 2e^2\sin(0.5) + e^2\cos(0.5) = 13.6.$$

Problems

37. Substituting $w = 65$ and $h = 160$, we have

$$f(65, 160) = 0.01(65^{0.25})(160^{0.75}) = 1.277 \text{ m}^2.$$

This tells us that a person who weighs 65 kg and is 160 cm tall has a surface area of about 1.277 m^2. Since

$$f_w(w, h) = 0.01(0.25w^{-0.75})h^{0.75} \text{ m}^2/\text{kg},$$

we have $f_w(65, 160) = 0.005$ m^2/kg. Thus, an increase of 1 kg in weight increases surface area by about 0.005 m^2. Since

$$f_h(w, h) = 0.01w^{0.25}(0.75h^{-0.25}) \text{ m}^2/\text{cm},$$

we have $f_h(65, 160) = 0.006$ m^2/cm. Thus, an increase of 1 cm in height increases surface area by about 0.006 m^2.

41. We compute the partial derivatives:

$$\frac{\partial Q}{\partial K} = b\alpha K^{\alpha-1}L^{1-\alpha} \quad \text{so} \quad K\frac{\partial Q}{\partial K} = b\alpha K^{\alpha}L^{1-\alpha}$$

$$\frac{\partial Q}{\partial L} = b(1-\alpha)K^{\alpha}L^{-\alpha} \quad \text{so} \quad L\frac{\partial Q}{\partial L} = b(1-\alpha)K^{\alpha}L^{1-\alpha}$$

Adding these two results, we have:

$$K\frac{\partial Q}{\partial K} + L\frac{\partial Q}{\partial L} = b(\alpha + 1 - \alpha)K^{\alpha}L^{1-\alpha} = Q.$$

Solutions for Section 14.3

Exercises

1. We have

$$z = e^y + x + x^2 + 6.$$

The partial derivatives are

$$\left.\frac{\partial z}{\partial x}\right|_{(x,y)=(1,0)} = (2x+1)\Big|_{(x,y)=(1,0)} = 3$$

$$\left.\frac{\partial z}{\partial y}\right|_{(x,y)=(1,0)} = e^y\Big|_{(x,y)=(1,0)} = 1.$$

So the equation of the tangent plane is

$$z = 9 + 3(x-1) + y = 6 + 3x + y.$$

5. Since $z_x = -e^{-x}\cos(y)$ and $z_y = -e^{-x}\sin(y)$, we have

$$dz = -e^{-x}\cos(y)dx - e^{-x}\sin(y)dy.$$

9. We have $dg = g_x\,dx + g_t\,dt$. Finding the partial derivatives, we have $g_x = 2x\sin(2t)$ so $g_x(2, \frac{\pi}{4}) = 4\sin(\pi/2) = 4$, and $g_t = 2x^2\cos(2t)$ so $g_t(2, \frac{\pi}{4}) = 8\cos(\frac{\pi}{2}) = 0$. Thus $dg = 4\,dx$.

Problems

13. **(a)** Since the equation of a tangent plane should be linear, this answer is wrong.
(b) The student didn't substitute the values $x = 2$, $y = 3$ into the formulas for the partial derivatives used in the formula of a tangent plane.
(c) Let $f(x, y) = z = x^3 - y^2$. Since $f_x(x, y) = 3x^2$ and $f_y(x, y) = -2y$, substituting $x = 2$, $y = 3$ gives $f_x(2, 3) = 12$ and $f_y(2, 3) = -6$. Then the equation of the tangent plane is

$$z = 12(x-2) - 6(y-3) - 1, \quad \text{or} \quad z = 12x - 6y - 7.$$

17. Making use of the values of P_r and P_L from the solution to Problem 17 on page 218, we have the local linearizations:
For $(r, L) = (8, 4000)$,

$$P(r, L) \approx 80 + 2.5(r-8) + 0.02(L-4000),$$

For $(r, L) = (8, 6000)$,

$$P(r, L) \approx 120 + 3.33(r-8) + 0.02(L-6000),$$

For $(r, L) = (13, 7000)$,

$$P(r, L) \approx 160 + 3.33(r-13) + 0.02(L-7000).$$

21. The error in η is approximated by $d\eta$, where

$$d\eta = \frac{\partial \eta}{\partial r}dr + \frac{\partial \eta}{\partial p}dp.$$

We need to find

$$\frac{\partial \eta}{\partial r} = \frac{\pi}{8}\frac{p4r^3}{v}$$

and

$$\frac{\partial \eta}{\partial p} = \frac{\pi}{8}\frac{r^4}{v}.$$

For $r = 0.005$ and $p = 10^5$ we get

$$\frac{\partial \eta}{\partial r}(0.005, 10^5) = 3.14159 \cdot 10^7, \quad \frac{\partial \eta}{\partial p}(0.005, 10^5) = 0.39270,$$

so that

$$d\eta = \frac{\partial \eta}{\partial r}dr + \frac{\partial \eta}{\partial p}dp.$$

is largest when we take all positive values to give

$$d\eta = 3.14159 \cdot 10^7 \cdot 0.00025 + 0.39270 \cdot 1000 = 8246.68.$$

This seems quite large but $\eta(0.005, 10^5) = 39269.9$ so the maximum error represents about 20% of any value computed by the given formula. Notice also the relative error in r is $\pm 5\%$, which means the relative error in r^4 is $\pm 20\%$.

25. **(a)** The area of a circle of radius r is given by

$$A = \pi r^2$$

and the perimeter is

$$L = 2\pi r.$$

Thus we get

$$r = \frac{L}{2\pi}$$

and

$$A = \pi\left(\frac{L}{2\pi}\right)^2 = \frac{L^2\pi}{4\pi^2} = \frac{L^2}{4\pi}.$$

Thus we get

$$\pi = \frac{L^2}{4A}.$$

(b) We will treat π as a function of L and A.

$$d\pi = \frac{\partial \pi}{\partial L} dL + \frac{\partial \pi}{\partial A} dA = \frac{2L}{4A} dL - \frac{L^2}{4A^2} dA.$$

If L is in error by a factor λ, then $\Delta L = \lambda L$, and if A is in error by a factor μ, then $\Delta A = \mu A$. Therefore,

$$\begin{aligned}
\Delta \pi &\approx \frac{2L}{4A}\Delta L - \frac{L^2}{4A^2}\Delta A \\
&= \frac{2L}{4A}\lambda L - \frac{L^2}{4A^2}\mu A \\
&= \frac{2\lambda L^2}{4A} - \frac{\mu L^2}{4A} = (2\lambda - \mu)\frac{L^2}{4A} = (2\lambda - \mu)\pi,
\end{aligned}$$

so π is in error by a factor of $2\lambda - \mu$.

Solutions for Section 14.4

Exercises

1. Since the partial derivatives are

$$\begin{aligned}
\frac{\partial f}{\partial x} &= \frac{15}{2}x^4 - 0 = \frac{15}{2}x^4 \\
\frac{\partial f}{\partial y} &= 0 - \frac{24}{7}y^5 = -\frac{24}{7}y^5
\end{aligned}$$

we have

$$\text{grad}\, f = \frac{\partial f}{\partial x}\vec{i} + \frac{\partial f}{\partial y}\vec{j} = \left(\frac{15}{2}x^4\right)\vec{i} - \left(\frac{24}{7}y^5\right)\vec{j}.$$

5. Since the partial derivatives are

$$f_x = \frac{x}{\sqrt{x^2+y^2}} \quad \text{and} \quad f_y = \frac{y}{\sqrt{x^2+y^2}},$$

we have

$$\nabla f = \frac{1}{\sqrt{x^2+y^2}}(x\vec{i} + y\vec{j}).$$

9. Since the partial derivatives are

$$z_x = (2x)\cos(x^2+y^2), \quad \text{and} \quad z_y = (2y)\cos(x^2+y^2),$$
$$\nabla z = 2x\cos(x^2+y^2)\vec{i} + 2y\cos(x^2+y^2)\vec{j}.$$

13. Since the partial derivatives are

$$\begin{aligned}
z_x &= \frac{e^y(x+y) - xe^y}{(x+y)^2} = \frac{ye^y}{(x+y)^2} \\
z_y &= \frac{xe^y(x+y) - xe^y}{(x+y)^2} = \frac{e^y(x^2+xy-x)}{(x+y)^2}
\end{aligned}$$

we have

$$\nabla z = \frac{ye^y}{(x+y)^2}\vec{i} + \frac{e^y(x^2+xy-x)}{(x+y)^2}\vec{j}$$

17. Since the partial derivatives are

$$f_r = 2\pi(h+r) \quad \text{and} \quad f_h = 2\pi r,$$

we have

$$\nabla f(2,3) = 10\pi\vec{i} + 4\pi\vec{j}.$$

21. Since the values of z decrease as we move in direction $\vec{i}$ from the point $(-2, 2)$, the directional derivative is negative.

25. Since the values of z decrease as we move in direction $\vec{i} + 2\vec{j}$ from the point $(0, -2)$, the directional derivative is negative.

29. The approximate direction of the gradient vector at point $(-2, 0)$ is $-\vec{i}$, since the gradient vector is perpendicular to the contour and points in the direction of increasing z-values. Answers may vary since answers are approximate and any positive multiple of the vector given is also correct.

33. The approximate direction of the gradient vector at point $(-2, 2)$ is $-\vec{i} + \vec{j}$, since the gradient vector is perpendicular to the contour and points in the direction of increasing z-values. Answers may vary since answers are approximate and any positive multiple of the vector given is also correct.

37. Since $f_x = 2\cos(2x - y)$ and $f_y = -\cos(2x - y)$, at $(1, 2)$ we have $\text{grad } f = 2\cos(0)\vec{i} - \cos(0)\vec{j} = 2\vec{i} - \vec{j}$. Thus

$$f_{\vec{u}}(1, 2) = \text{grad } f \cdot \left(\frac{3}{5}\vec{i} - \frac{4}{5}\vec{j}\right) = \frac{2 \cdot 3 - 1(-4)}{5} = \frac{10}{5} = 2.$$

Problems

41. Since $\vec{u} = (\vec{i} - \vec{j})/\sqrt{2}$, we head away from the point $(3, 1)$ toward the point $(4, 0)$.
From the graph, we see that $f(3, 1) = 1$ and $f(4, 0) = 4$. Since the points $(3, 1)$ and $(4, 0)$ are a distance $\sqrt{2}$ apart, we have

$$f_{\vec{u}}(3, 1) \approx \frac{f(4, 0) - f(3, 1)}{\sqrt{2}} = \frac{4 - 1}{\sqrt{2}} = 2.12.$$

45. The directional derivative is approximately the change in z (as we move in direction $\vec{v}$) divided by the horizontal change in position. In the direction of $\vec{v}$, the directional derivative $\approx \dfrac{2 - 1}{\|\vec{i} + \vec{j}\|} = \dfrac{1}{\sqrt{2}} \approx 0.7$.

49. **(a)** In the $\vec{i} - \vec{j}$ direction the function is decreasing, so the value of $g_{\vec{u}}(2, 5)$ is negative.
(b) In the $\vec{i} + \vec{j}$ direction the function is decreasing, so the value of $g_{\vec{u}}(2, 5)$ negative as well.

53. **(a)** First we will find a unit vector in the same direction as the vector $\vec{v} = 3\vec{i} - 2\vec{j}$. Since this vector has magnitude $\sqrt{13}$, a unit vector is

$$\vec{u}_1 = \frac{1}{\|\vec{v}\|}\vec{v} = \frac{3}{\sqrt{13}}\vec{i} - \frac{2}{\sqrt{13}}\vec{j}.$$

The partial derivatives are

$$f_x(x, y) = \frac{(1 + x^2) - (x + y) \cdot 2x}{(1 + x^2)^2} = \frac{1 - x^2 - 2xy}{(1 + x^2)^2},$$

$$\text{and} \quad f_y(x, y) = \frac{1}{1 + x^2},$$

then, at the point P, we have

$$f_x(P) = f_x(1, -2) = \frac{1 - 1^2 - 2 \cdot 1 \cdot (-2)}{(1 + 1^2)^2} = 1,$$

$$f_y(P) = f_y(1, -2) = f_y(1, -2) = \frac{1}{1 + 1^2} = \frac{1}{2}.$$

Thus

$$\begin{aligned} f_{\vec{u}_1}(P) &= \text{grad } f(P) \cdot \vec{u}_1 \\ &= (\vec{i} + \frac{1}{2}\vec{j}) \cdot (\frac{3}{\sqrt{13}}\vec{i} - \frac{2}{\sqrt{13}}\vec{j}) \\ &= \frac{3}{\sqrt{13}} - \frac{1}{\sqrt{13}} = \frac{2}{\sqrt{13}}. \end{aligned}$$

(b) The unit vector in the same direction as the vector $\vec{v} = -\vec{i} + 4\vec{j}$ is

$$\begin{aligned}\vec{u}_2 &= \frac{1}{\|\vec{v}\|}\vec{v} = \frac{1}{\sqrt{(-1)^2+4^2}}(-\vec{i}+4\vec{j}) \\ &= -\frac{1}{\sqrt{17}}\vec{i} + \frac{4}{\sqrt{17}}\vec{j}.\end{aligned}$$

Since we have calculated from part (a) that $f_x(P) = 1$ and $f_y(P) = 1/2$,

$$\begin{aligned}f_{\vec{u}_2}(P) &= \text{grad}\, f(P) \cdot \vec{u}_2 \\ &= (\vec{i} + \frac{1}{2}\vec{j}) \cdot (-\frac{1}{\sqrt{17}}\vec{i} + \frac{4}{\sqrt{17}}\vec{j}) \\ &= -\frac{1}{\sqrt{17}} + \frac{2}{\sqrt{17}} = \frac{1}{\sqrt{17}}.\end{aligned}$$

(c) The direction of greatest increase is grad f at P. By part (a) we have found that

$$f_x(P) = 1 \quad \text{and} \quad f_y(P) = \frac{1}{2}.$$

Therefore the direction of greatest increase is

$$\text{grad}\, f(P) = \vec{i} + \frac{1}{2}\vec{j}.$$

57. Let's put a coordinate plane on the area you are hiking, with your trail along the x-axis and the second trail branching off at the origin as in Figure 14.1. You are moving in the positive x direction. Let $h(x, y)$ be the elevation at the point (x, y) on the mountain.

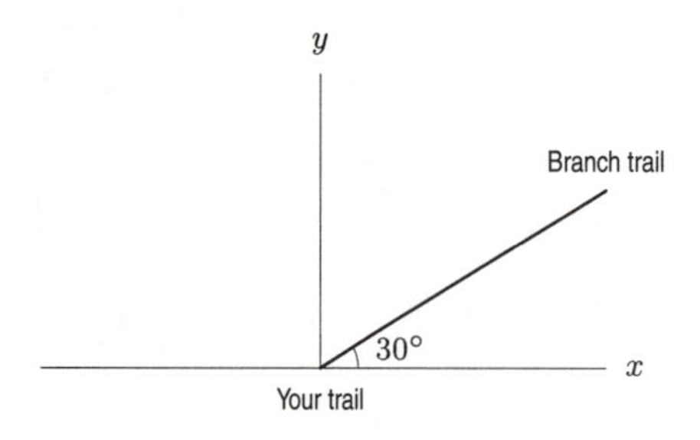

Figure 14.1: Two trails

Since the trail along the x-axis ascends at a 20° angle, we have $h_x(0,0) = \tan 20^\circ$. Since the trail is the steepest path, grad h must point along your trail in the positive x direction. Thus

$$\text{grad}\, h = h_x\vec{i} + 0\vec{j} = \tan 20^\circ \vec{i}.$$

We must compute the rate of change of elevation in the direction of the branch trail. The unit vector in this direction is $\vec{u} = \cos 30^\circ \vec{i} + \sin 30^\circ \vec{j}$, and thus the directional derivative is

$$h_{\vec{u}} = (\text{grad}\, h) \cdot \vec{u} = (\tan 20^\circ \vec{i}) \cdot (\cos 30^\circ \vec{i} + \sin 30^\circ \vec{j}) = (\tan 20^\circ)(\cos 30^\circ) = 0.3152.$$

The angle of ascent of the branch trail is thus $\tan^{-1}(0.3152) = 17.5^\circ$.

61. First, we check that $(2)(3) = 6$. Then let $f(x, y) = xy$ so that the given curve is the contour $f(x, y) = 6$. Since $f_x = y$ and $f_y = x$, we have grad $f(2,3) = 3\vec{i} + 2\vec{j}$. Since gradients are perpendicular to contours, a vector normal to the curve at $(2,3)$ is $\vec{n} = 3\vec{i} + 2\vec{j}$. Using the normal vector to a line the same way we use the normal vector to a plane, we get that the equation of the tangent line is $3(x-2) + 2(y-3) = 0$.

65. At the point $(1.2, 0)$, the value of the function is 4.2. Nearby, the largest value is 8.9 at the point $(1.4, -1)$. Since the gradient vector points in the direction of maximum increase, it points into the fourth quadrant.

69. **(a)** P corresponds to greatest rate of increase of f and Q corresponds to greatest rate of decrease of f. See Figure 14.2.
(b) The points are marked in Figure 14.3.
(c) Amplitude is $\|\operatorname{grad} f\|$. The equation is

$$f_{\vec{u}} = \|\operatorname{grad} f\| \cos\theta.$$

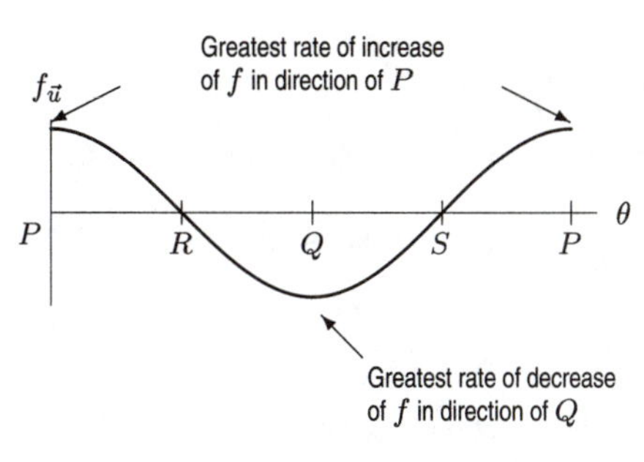

Figure 14.2

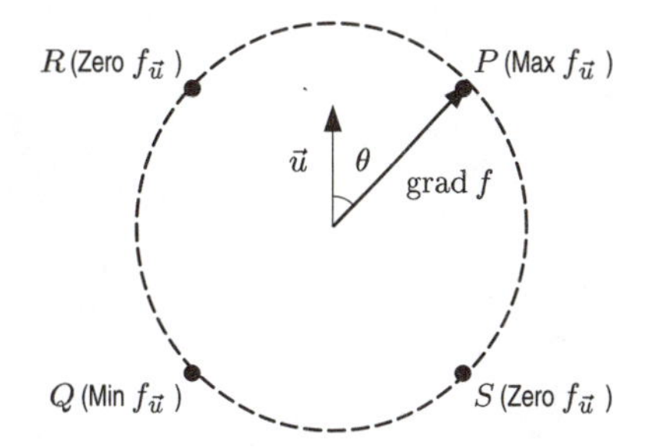

Figure 14.3

Solutions for Section 14.5

Exercises

1. We have $f_x = 2x$, $f_y = 3y^2$, and $f_z = -4z^3$. Thus

$$\operatorname{grad} f = 2x\vec{i} + 3y^2\vec{j} - 4z^3\vec{k}.$$

5. We have $f_x = -2x/(x^2+y^2+z^2)^2$, $f_y = -2y/(x^2+y^2+z^2)^2$, and $f_z = -2z/(x^2+y^2+z^2)^2$. Thus

$$\operatorname{grad} f = \frac{-2}{(x^2+y^2+z^2)^2}(x\vec{i} + y\vec{j} + z\vec{k}).$$

9. We have $f_x = yz$, $f_y = xz$, and $f_z = yz$. Thus $f_x(1,2,3) = 6$, $f_y(1,2,3) = 3$, and $f_z(1,2,3) = 2$, so

$$\operatorname{grad} f = 6\vec{i} + 3\vec{j} + 2\vec{k}.$$

13. We have $\operatorname{grad} f = y\vec{i} + x\vec{j} + 2z\vec{k}$, so $\operatorname{grad} f(1,1,0) = \vec{i} + \vec{j}$. A unit vector in the direction we want is $u = (1/\sqrt{2})(-\vec{i} + \vec{k})$. Therefore, the directional derivative is

$$\operatorname{grad} f(1,1,0)\cdot\vec{u} = \frac{1(-1) + 1\cdot 0 + 0\cdot 1}{\sqrt{2}} = \frac{-1}{\sqrt{2}}.$$

17. First, we check that $(-1)^2 - (1)^2 + 2^2 = 4$. Then let $f(x,y,z) = x^2 - y^2 + z^2$ so that the given surface is the level surface $f(x,y,z) = 4$. Since $f_x = 2x$, $f_y = -2y$, and $f_z = 2z$, we have $\operatorname{grad} f(-1,1,2) = -2\vec{i} - 2\vec{j} + 4\vec{k}$. Since gradients are perpendicular to level surfaces, a vector normal to the surface at $(-1,1,2)$ is $\vec{n} = -2\vec{i} - 2\vec{j} + 4\vec{k}$. Thus an equation for the tangent plane is

$$-2(x+1) - 2(y-1) + 4(z-2) = 0.$$

21. First, we check that $1 = 2^2 - 3$. Then we let $f(x,y,z) = y^2 - z^2 + 3$, so that the given surface is the level surface $f(x,y,z) = 0$. Since $f_x = 0$, $f_y = 2y$, and $f_z = -2z$, we have $\operatorname{grad} f(-1,1,2) = 2\vec{j} - 4\vec{k}$. Since gradients are perpendicular to level surfaces, a vector normal to the surface at $(-1,1,2)$ is $\vec{n} = 2\vec{j} - 4\vec{k}$. Thus an equation for the tangent plane is

$$2(y-1) - 4(z-2) = 0.$$

Problems

25. (a) The function $T(x, y, z) =$ constant where $x^2 + y^2 + z^2 =$ constant. These surfaces are spheres centered at the origin.

(b) Calculating the partial derivative with respect to x gives

$$\frac{\partial T}{\partial x} = -2xe^{-(x^2+y^2+z^2)}.$$

Similar calculations for the other variables shows that

$$\operatorname{grad} T = (-2x\vec{i} - 2y\vec{j} - 2z\vec{k})e^{-(x^2+y^2+z^2)}.$$

(c) At the point $(1, 0, 0)$

$$\operatorname{grad} T(1, 0, 0) = -2e^{-1}\vec{i}.$$

Moving from the point $(1, 0, 0)$ to $(2, 1, 0)$, you move in the direction

$$(2 - 1)\vec{i} + (1 - 0)\vec{j} = \vec{i} + \vec{j}.$$

A unit vector in this direction is

$$\vec{u} = \frac{\vec{i} + \vec{j}}{\sqrt{2}}.$$

The directional derivative of $T(x, y, z)$ in this direction at the point $(1, 0, 0)$ is

$$T_{\vec{u}}(1, 0, 0) = -2e^{-1}\vec{i} \cdot \frac{\vec{i} + \vec{j}}{\sqrt{2}} = -\sqrt{2}e^{-1}.$$

Since you are moving at a speed of 3 units per second,

$$\text{Rate of change of temperature } = -\sqrt{2}e^{-1} \cdot 3 = -3\sqrt{2}e^{-1} \text{ degrees/second.}$$

29. **(a)** The surface is the level surface $F(x, y, z) = 7$, where $F(x, y, z) = x^2 + y^2 - xyz$. Thus the normal vector to the tangent plane is $\operatorname{grad} F = (2x - yz)\vec{i} + (2y - xz)\vec{j} + (-xy)\vec{k}$. Evaluated at $(2, 3, 1)$, we get the normal to the plane

$$\vec{n} = \vec{i} + 4\vec{j} - 6\vec{k}.$$

Thus the equation of the plane is

$$(x - 2) + 4(y - 3) - 6(z - 1) = 0.$$

(b) Solving $x^2 + y^2 - xyz = 7$ for z, we get

$$z = \frac{x^2 + y^2 - 7}{xy}.$$

Thus, we have

$$f(x, y) = \frac{x^2 + y^2 - 7}{xy} = \frac{x}{y} + \frac{y}{x} - \frac{7}{xy}.$$

We have

$$f_x(x, y) = \frac{1}{y} - \frac{y}{x^2} + \frac{7}{x^2y}$$

$$f_y(x, y) = -\frac{x}{y^2} + \frac{1}{x} + \frac{7}{xy^2}.$$

Thus $f_x(2, 3) = 1/3 - 3/4 + 7/12 = 1/6$ and $f_y(2, 3) = -2/9 + 1/2 + 7/18 = 2/3$. Thus the equation of the tangent plane is

$$z = 1 + (1/6)(x - 2) + (2/3)(y - 3).$$

This is the same as the answer to part (a) when that equation is solved for z.

33. If write $\vec{r} = x\vec{i} + y\vec{j} + z\vec{k}$, then we know

$$\operatorname{grad} f(x, y, z) = g(x, y, z)(x\vec{i} + y\vec{j} + z\vec{k}) = g(x, y, z)\vec{r}$$

so $\operatorname{grad} f$ is everywhere radially outward, and therefore perpendicular to a sphere centered at the origin. If f were not constant on such a sphere, then $\operatorname{grad} f$ would have a component tangent to the sphere. Thus, f must be constant on any sphere centered at the origin.

Solutions for Section 14.6

Exercises

1. Using the chain rule we see:

$$\begin{aligned}\frac{dz}{dt} &= \frac{\partial z}{\partial x}\frac{dx}{dt} + \frac{\partial z}{\partial y}\frac{dy}{dt} \\ &= -y^2 e^{-t} + 2xy\cos t \\ &= -(\sin t)^2 e^{-t} + 2e^{-t}\sin t\cos t \\ &= \sin(t)e^{-t}(2\cos t - \sin t)\end{aligned}$$

We can also solve the problem using one variable methods:

$$\begin{aligned}z &= e^{-t}(\sin t)^2 \\ \frac{dz}{dt} &= \frac{d}{dt}(e^{-t}(\sin t)^2) \\ &= \frac{de^{-t}}{dt}(\sin t)^2 + e^{-t}\frac{d(\sin t)^2}{dt} \\ &= -e^{-t}(\sin t)^2 + 2e^{-t}\sin t\cos t \\ &= e^{-t}\sin t(2\cos t - \sin t)\end{aligned}$$

5. Substituting into the chain rule gives

$$\begin{aligned}\frac{dz}{dt} &= \frac{\partial z}{\partial x}\frac{dx}{dt} + \frac{\partial z}{\partial y}\frac{dy}{dt} = e^y(2) + xe^y(-2t) \\ &= 2e^y(1 - xt) = 2e^{1-t^2}(1 - 2t^2).\end{aligned}$$

9. Since z is a function of two variables x and y which are functions of two variables u and v, the two chain rule identities which apply are:

$$\begin{aligned}\frac{\partial z}{\partial u} &= \frac{\partial z}{\partial x}\frac{\partial x}{\partial u} + \frac{\partial z}{\partial y}\frac{\partial y}{\partial u} = e^y\left(\frac{1}{u}\right) + xe^y \cdot 0 = \frac{e^v}{u}. \\ \frac{\partial z}{\partial v} &= \frac{\partial z}{\partial x}\frac{\partial x}{\partial v} + \frac{\partial z}{\partial y}\frac{\partial y}{\partial v} = e^y(0) + xe^y \cdot 1 = e^v \ln u.\end{aligned}$$

13. Since z is a function of two variables x and y which are functions of two variables u and v, the two chain rule identities which apply are:

$$\begin{aligned}\frac{\partial z}{\partial u} &= \frac{\partial z}{\partial x}\frac{\partial x}{\partial u} + \frac{\partial z}{\partial y}\frac{\partial y}{\partial u} = \left(\cos\left(\frac{x}{y}\right)\right)\left(\frac{1}{y}\right)\frac{1}{u} + \left(\cos\left(\frac{x}{y}\right)\right)\left(\frac{-x}{y^2}\right)\cdot 0 \\ &= \frac{1}{vu}\cos\left(\frac{\ln u}{v}\right). \\ \frac{\partial z}{\partial v} &= \frac{\partial z}{\partial x}\frac{\partial x}{\partial v} + \frac{\partial z}{\partial y}\frac{\partial y}{\partial v} = \left(\cos\left(\frac{x}{y}\right)\right)\left(\frac{1}{y}\right)\cdot 0 + \left(\cos\left(\frac{x}{y}\right)\right)\left(\frac{-x}{y^2}\right)\cdot 1 = -\frac{\ln u}{v^2}\cos\left(\frac{\ln u}{v}\right).\end{aligned}$$

Problems

17.

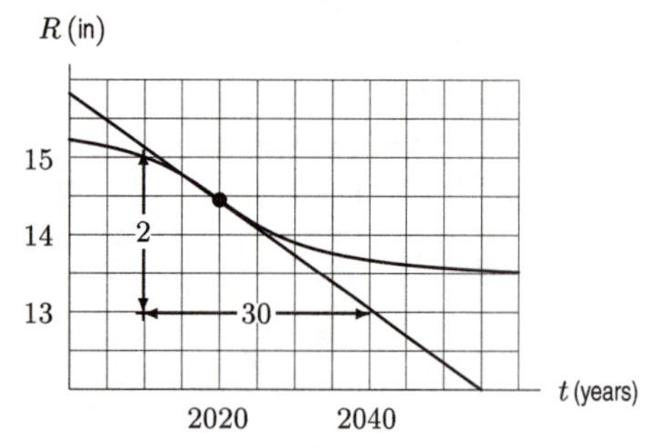

Figure 14.4: Global warming predictions: Rainfall as a function of time

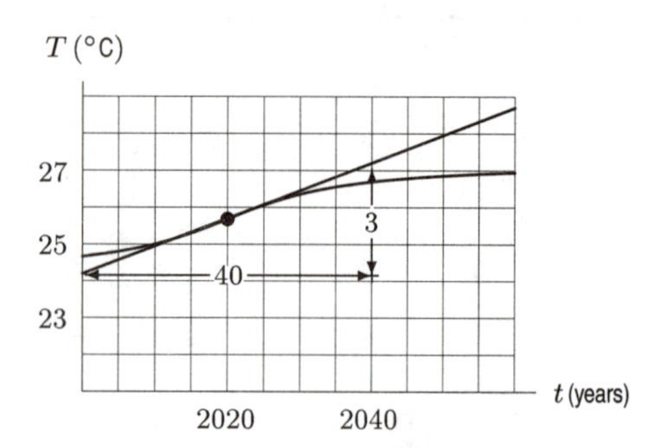

Figure 14.5: Global warming predictions: Temperature as a function of time

We know that, as long as the temperature and rainfall stay close to their current values of $R = 15$ inches and $T = 30°$C, a change, ΔR, in rainfall and a change, ΔT, in temperature produces a change, ΔC, in corn production given by

$$\Delta C \approx 3.3\Delta R - 5\Delta T.$$

Now both R and T are functions of time t (in years), and we want to find the effect of a small change in time, Δt, on R and T. Figure 14.4 shows that the slope of the graph for R versus t is about $-2/30 \approx -0.07$ in/year when $t = 2020$. Similarly, Figure 14.5 shows the slope of the graph of T versus t is about $3/40 \approx 0.08°$C/year when $t = 2020$. Thus, around the year 2020,

$$\Delta R \approx -0.07\Delta t \quad \text{and} \quad \Delta T \approx 0.08\Delta t.$$

Substituting these into the equation for ΔC, we get

$$\Delta C \approx (3.3)(-0.07)\Delta t - (5)(0.08)\Delta t \approx -0.6\Delta t.$$

Since at present $C = 100$, corn production will decline by about 0.6 % between the years 2020 and 2021. Now $\Delta C \approx -0.6\Delta t$ tells us that when $t = 2020$,

$$\frac{\Delta C}{\Delta t} \approx -0.6, \quad \text{and therefore, that} \quad \frac{dC}{dt} \approx -0.6.$$

21. All are done using the chain rule.

(a) We have $u = x$, $v = 3$. Thus $du/dx = 1$ and $dv/dx = 0$ so

$$f'(x) = F_u(x,3)(1) + F_v(x,3)(0) = F_u(x,3).$$

(b) We have $u = 3$, $v = x$. Thus $du/dx = 0$ and $dv/dx = 1$ so

$$f'(x) = F_u(3,x)(0) + F_v(3,x)(1) = F_v(3,x).$$

(c) We have $u = x$, $v = x$. Thus $du/dx = dv/dx = 1$ so

$$f'(x) = F_u(x,x)(1) + F_v(x,x)(1) = F_u(x,x) + F_v(x,x).$$

(d) We have $u = 5x$, $v = x^2$. Thus $du/dx = 5$ and $dv/dx = 2x$ so

$$f'(x) = F_u(5x,x^2)(5) + F_v(5x,x^2)(2x).$$

25. Since $\left(\frac{\partial U}{\partial P}\right)_V$ involves the variables P and V, we are viewing U as a function of these two variables, so $U = U_3(P,V)$. Then

$$\left(\frac{\partial U}{\partial P}\right)_V = \frac{\partial U_3(P,V)}{\partial P}.$$

29. We will use analysis similar to that in Example 7. Since V is a function of P and T, we have

$$dV = \left(\frac{\partial V}{\partial T}\right)_P dT + \left(\frac{\partial V}{\partial P}\right)_T dP$$

We are interested in $\left(\frac{\partial U}{\partial V}\right)_T$ so we use the formula for dU corresponding to U_2. Substituting g for dV into this formula for dU gives

$$\begin{aligned} dU &= \left(\frac{\partial U}{\partial T}\right)_V dT + \left(\frac{\partial U}{\partial V}\right)_T \left(\left(\frac{\partial V}{\partial T}\right)_P dT + \left(\frac{\partial V}{\partial P}\right)_T dP\right) \\ &= \left(\left(\frac{\partial U}{\partial T}\right)_V + \left(\frac{\partial U}{\partial V}\right)_T \left(\frac{\partial V}{\partial T}\right)_P\right) dT + \left(\frac{\partial U}{\partial V}\right)_T \left(\frac{\partial V}{\partial P}\right)_T dP \end{aligned}$$

But we are also interested in $\left(\frac{\partial U}{\partial P}\right)_T$ so we compare with the formula for dU corresponding to U_1.

$$dU = \left(\frac{\partial U}{\partial T}\right)_P dT + \left(\frac{\partial U}{\partial P}\right)_T dP.$$

Since the coefficients of dP must be identical, we get

$$\left(\frac{\partial U}{\partial P}\right)_T = \left(\frac{\partial U}{\partial V}\right)_T \left(\frac{\partial V}{\partial P}\right)_T.$$

33. Use chain rule for the equation $0 = F(x, y, f(x, y))$. Differentiating both sides with respect to x, remembering $z = f(x, y)$ and regarding y as a constant gives:

$$0 = \frac{\partial F}{\partial x}\frac{dx}{dx} + \frac{\partial F}{\partial z}\frac{dz}{dx}.$$

Since $dx/dx = 1$, we get

$$-\frac{\partial F}{\partial x} = \frac{\partial F}{\partial z}\frac{\partial z}{\partial x},$$

so

$$\frac{\partial z}{\partial x} = -\frac{\partial F/\partial x}{\partial F/\partial z}.$$

Similarly, differentiating both sides of the equation $0 = F(x, y, f(x, y))$ with respect to y gives:

$$0 = \frac{\partial F}{\partial y}\frac{dy}{dy} + \frac{\partial F}{\partial z}\frac{dz}{dy}.$$

Since $dy/dy = 1$, we get

$$-\frac{\partial F}{\partial y} = \frac{\partial F}{\partial z}\frac{\partial z}{\partial y},$$

so

$$\frac{\partial z}{\partial y} = -\frac{\partial F/\partial y}{\partial F/\partial z}.$$

Solutions for Section 14.7

Exercises

1. We have $f_x = 6xy + 5y^3$ and $f_y = 3x^2 + 15xy^2$, so $f_{xx} = 6y$, $f_{xy} = 6x + 15y^2$, $f_{yx} = 6x + 15y^2$, and $f_{yy} = 30xy$.

5. Since $f = (x + y)e^y$, the partial derivatives are

$$\begin{aligned} f_x &= e^y, \quad f_y = e^y(x + 1 + y) \\ f_{xx} &= 0, \quad f_{yx} = e^y = f_{xy} \\ f_{yy} &= xe^y + e^y + e^y + ye^y = e^y(x + 2 + y). \end{aligned}$$

9. We have $f_x = 6\cos 2x \cos 5y$ and $f_y = -15\sin 2x \sin 5y$, so $f_{xx} = -12\sin 2x \cos 5y$, $f_{xy} = -30\cos 2x \sin 5y$, $f_{yx} = -30\cos 2x \sin 5y$, and $f_{yy} = -75\sin 2x \cos 5y$.

13. The quadratic Taylor expansion about $(0,0)$ is given by

$$f(x,y) \approx Q(x,y) = f(0,0) + f_x(0,0)x + f_y(0,0)y + \frac{1}{2}f_{xx}(0,0)x^2 + f_{xy}(0,0)xy + \frac{1}{2}f_{yy}(0,0)y^2.$$

First we find all the relevant derivatives

$$\begin{aligned} f(x,y) &= \cos(x+3y) \\ f_x(x,y) &= -\sin(x+3y) \\ f_y(x,y) &= -3\sin(x+3y) \\ f_{xx}(x,y) &= -\cos(x+3y) \\ f_{yy}(x,y) &= -9\cos(x+3y) \\ f_{xy}(x,y) &= -3\cos(x+3y) \end{aligned}$$

Now we evaluate each of these derivatives at $(0,0)$ and substitute into the formula to get as our final answer:

$$Q(x,y) = 1 - \frac{1}{2}x^2 - 3xy - \frac{9}{2}y^2$$

Notice this is the same as what you get if you substitute $x+3y$ for u in the single variable quadratic approximation $Q(u) = 1 - u^2/2$ for $\cos u$.

17. The quadratic Taylor expansion about $(0,0)$ is given by

$$f(x,y) \approx Q(x,y) = f(0,0) + f_x(0,0)x + f_y(0,0)y + \frac{1}{2}f_{xx}(0,0)x^2 + f_{xy}(0,0)xy + \frac{1}{2}f_{yy}(0,0)y^2$$

So first we find all the relevant derivatives

$$\begin{aligned} f(x,y) &= \sin 2x + \cos y \\ f_x(x,y) &= 2\cos 2x \\ f_y(x,y) &= -\sin y \\ f_{xx}(x,y) &= -4\sin 2x \\ f_{yy}(x,y) &= -\cos y \\ f_{xy}(x,y) &= 0 \end{aligned}$$

We substitute into the formula to get for our answer:

$$Q(x,y) = 1 + 2x - \frac{1}{2}y^2$$

21. **(a)** $f_x(P) < 0$ because f decreases as you go to the right.
(b) $f_y(P) = 0$ because f does not change as you go up.
(c) $f_{xx}(P) < 0$ because f_x decreases as you go to the right (f_x changes from a small negative number to a large negative number).
(d) $f_{yy}(P) = 0$ because f_y does not change as you go up.
(e) $f_{xy}(P) = 0$ because f_x does not change as you go up.

25. **(a)** $f_x(P) < 0$ because f decreases as you go to the right.
(b) $f_y(P) < 0$ because f decreases as you go up.
(c) $f_{xx}(P) = 0$ because f_x does not change as you go to the right. (Notice that the level curves are equidistant and parallel, so the partial derivatives of f do not change if you move horizontally or vertically.)
(d) $f_{yy}(P) = 0$ because f_y does not change as you go up.
(e) $f_{xy}(P) = 0$ because f_x does not change as you go up.

29. We have $f(1,0) = 1$ and the relevant derivatives are:

$$\begin{aligned} f_x &= \frac{1}{2}(x+2y)^{-1/2} \quad \text{so} \quad f_x(1,0) = \frac{1}{2} \\ f_y &= (x+2y)^{-1/2} \quad \text{so} \quad f_y(1,0) = 1 \\ f_{xx} &= -\frac{1}{4}(x+2y)^{-3/2} \quad \text{so} \quad f_{xx}(1,0) = -\frac{1}{4} \\ f_{xy} &= -\frac{1}{2}(x+2y)^{-3/2} \quad \text{so} \quad f_{xy}(1,0) = -\frac{1}{2} \\ f_{yy} &= -(x+2y)^{-3/2} \quad \text{so} \quad f_{yy}(1,0) = -1\,. \end{aligned}$$

Thus the linear approximation, $L(x, y)$ to $f(x, y)$ at $(1, 0)$, is given by:

$$\begin{aligned} f(x,y) \approx L(x,y) &= f(1,0) + f_x(1,0)(x-1) + f_y(1,0)(y-0) \\ &= 1 + \frac{1}{2}(x-1) + y\,. \end{aligned}$$

The quadratic approximation, $Q(x, y)$ to $f(x, y)$ near $(1, 0)$, is given by:

$$\begin{aligned} f(x,y) \approx Q(x,y) &= f(1,0) + f_x(1,0)(x-1) + f_y(1,0)(y-0) + \frac{1}{2}f_{xx}(1,0)(x-1)^2 \\ &\quad + f_{xy}(1,0)(x-1)(y-0) + \frac{1}{2}f_{yy}(1,0)(y-0)^2 \\ &= 1 + \frac{1}{2}(x-1) + y - \frac{1}{8}(x-1)^2 - \frac{1}{2}(x-1)y - \frac{1}{2}y^2\,. \end{aligned}$$

The values of the approximations are

$$\begin{aligned} L(0.9, 0.2) &= 1 - 0.05 + 0.2 = 1.15 \\ Q(0.9, 0.2) &= 1 - 0.05 + 0.2 - 0.00125 + 0.01 - 0.02 = 1.13875 \end{aligned}$$

and the exact value is

$$f(0.9, 0.2) = \sqrt{1.3} \approx 1.14018.$$

Observe that the quadratic approximation is closer to the exact value.

Problems

33. **(a)** $z_{yx} = z_{xy} = 4y$

(b) $z_{xyx} = \dfrac{\partial}{\partial x}(z_{xy}) = \dfrac{\partial}{\partial x}(4y) = 0$

(c) $z_{xyy} = z_{yxy} = \dfrac{\partial}{\partial y}(4y) = 4$

37. **(a)** Calculate the partial derivatives:

$$\begin{array}{lll} f(x,y) = \sin x \sin y & f(0,0) = 0 & f(\frac{\pi}{2},\frac{\pi}{2}) = 1 \\ f_x(x,y) = \cos x \sin y & f_x(0,0) = 0 & f_x(\frac{\pi}{2},\frac{\pi}{2}) = 0 \\ f_y(x,y) = \sin x \cos y & f_y(0,0) = 0 & f_y(\frac{\pi}{2},\frac{\pi}{2}) = 0 \\ f_{xx}(x,y) = -\sin x \sin y & f_{xx}(0,0) = 0 & f_{xx}(\frac{\pi}{2},\frac{\pi}{2}) = -1 \\ f_{xy}(x,y) = \cos x \cos y & f_{xy}(0,0) = 1 & f_{xy}(\frac{\pi}{2},\frac{\pi}{2}) = 0 \\ f_{yy}(x,y) = -\sin x \sin y & f_{yy}(0,0) = 0 & f_{yy}(\frac{\pi}{2},\frac{\pi}{2}) = -1 \end{array}$$

Thus, the Taylor polynomial about $(0, 0)$ is

$$f(x,y) \approx Q_1(x,y) = xy.$$

The Taylor polynomial about $(\frac{\pi}{2}, \frac{\pi}{2})$ is

$$f(x,y) \approx Q_2(x,y) = 1 - \frac{1}{2}\left(x - \frac{\pi}{2}\right)^2 - \frac{1}{2}\left(y - \frac{\pi}{2}\right)^2.$$

(b)

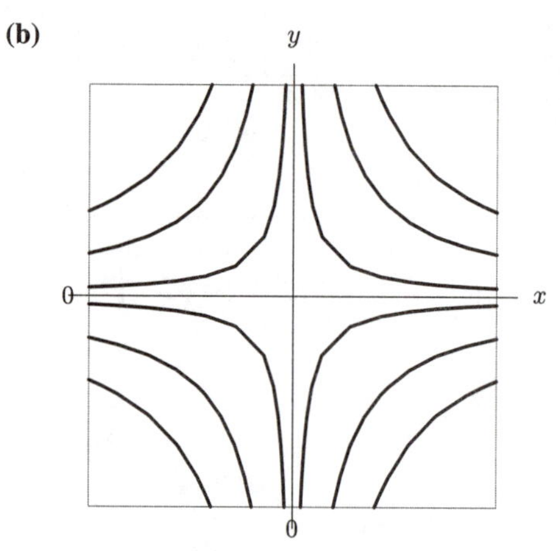

Figure 14.6: $f(x, y) \approx xy$: Quadratic approximation about $(0, 0)$

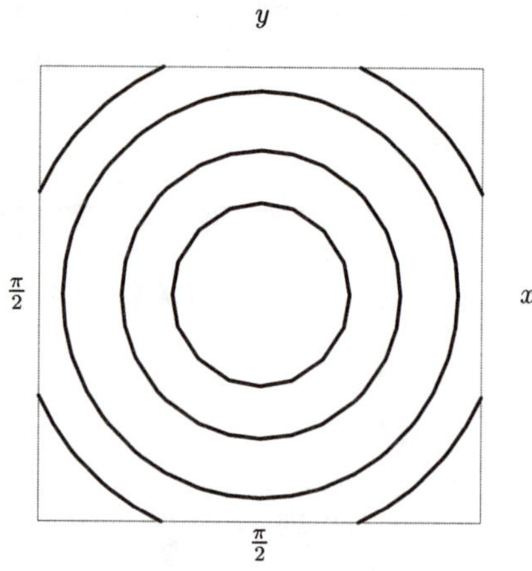

Figure 14.7: $f(x, y) \approx 1 - \frac{1}{2}(x - \frac{\pi}{2})^2 - \frac{1}{2}(y - \frac{\pi}{2})^2$: Quadratic approximation about $(\frac{\pi}{2}, \frac{\pi}{2})$

41. Letting $G(t) = f(t, 0)$ so that $G'(t) = f_x(t, 0)$, the Fundamental Theorem tells us that $\int_{t=0}^{a} G'(t)dt = G(a) - G(0)$. Thus

$$\int_{t=0}^{a} f_x(t, 0)dt = f(a, 0) - f(0, 0)$$

Letting $H(t) = f(a, t)$ so that $H'(t) = f_y(a, t)$, the Fundamental Theorem tells us that $\int_{t=0}^{b} H'(t)dt = H(b) - H(0)$. Thus

$$\int_{t=0}^{b} f_y(a, t)dt = f(a, b) - f(a, 0)$$

Thus

$$\begin{aligned} f(0, 0) + \int_{t=0}^{a} f_x(t, 0)dt + \int_{t=0}^{b} f_y(a, t)dt \\ = f(0, 0) + (f(a, 0) - f(0, 0)) + (f(a, b) - f(a, 0)) \\ = f(a, b). \end{aligned}$$

Solutions for Section 14.8

Exercises

1. Not differentiable at $(0, 0)$.

5. Not differentiable at all points on the x or y axes.

Problems

9. **(a)** The contour diagram for $f(x, y) = xy/\sqrt{x^2 + y^2}$ is shown in Figure 14.8.

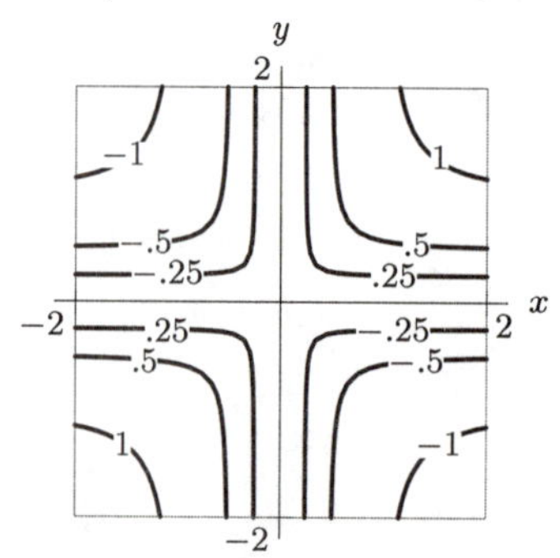

Figure 14.8

(b) By the chain rule, f is differentiable at all points (x, y) where $x^2 + y^2 \neq 0$, and so at all points $(x, y) \neq (0, 0)$.

(c) The partial derivatives of f are given by

$$f_x(x, y) = \frac{y^3}{(x^2 + y^2)^{3/2}}, \qquad \text{for} \quad (x, y) \neq (0, 0),$$

and

$$f_y(x, y) = \frac{x^3}{(x^2 + y^2)^{3/2}}, \qquad \text{for} \quad (x, y) \neq (0, 0).$$

Both f_x and f_y are continuous at $(x, y) \neq (0, 0)$.

(d) If f were differentiable at $(0, 0)$, the chain rule would imply that the function

$$g(t) = \begin{cases} f(t, t), & t \neq 0 \\ 0, & t = 0 \end{cases}$$

would be differentiable at $t = 0$. But

$$g(t) = \frac{t^2}{\sqrt{2t^2}} = \frac{1}{\sqrt{2}} \cdot \frac{t^2}{|t|} = \frac{1}{\sqrt{2}} \cdot |t|,$$

which is not differentiable at $t = 0$. Hence, f is not differentiable at $(0, 0)$.

(e) The partial derivatives of f at $(0, 0)$ are given by

$$f_x(0, 0) = \lim_{x \to 0} \frac{f(x, 0) - f(0, 0)}{x} = \lim_{x \to 0} \frac{\frac{x \cdot 0}{\sqrt{x^2 + 0^2}} - 0}{x} = \lim_{x \to 0} \frac{0 - 0}{x} = 0,$$

$$f_y(0, 0) = \lim_{y \to 0} \frac{f(0, y) - f(0, 0)}{y} = \lim_{y \to 0} \frac{\frac{0 \cdot y}{\sqrt{0^2 + y^2}} - 0}{y} = \lim_{y \to 0} \frac{0 - 0}{y} = 0.$$

The limit $\lim_{(x,y) \to (0,0)} f_x(x, y)$ doesn't exist since if we choose $x = y = t$, $t \neq 0$, then

$$f_x(x, y) = f_x(t, t) = \frac{t^3}{(2t^2)^{3/2}} = \frac{t^3}{2\sqrt{2} \cdot |t|^3} = \begin{cases} \frac{1}{2\sqrt{2}}, & t > 0, \\ -\frac{1}{2\sqrt{2}}, & t < 0. \end{cases}$$

Thus, f_x is not continuous at $(0, 0)$. Similarly, f_y is not continuous at $(0, 0)$.

13. (a) The contour diagram of $f(x, y) = \sqrt{|xy|}$ is shown in Figure 14.9.

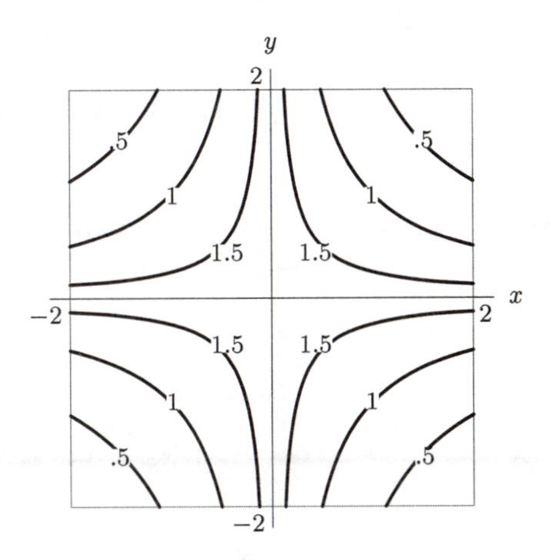

Figure 14.9

(b) The graph of $f(x, y) = \sqrt{|xy|}$ is shown in Figure 14.10.

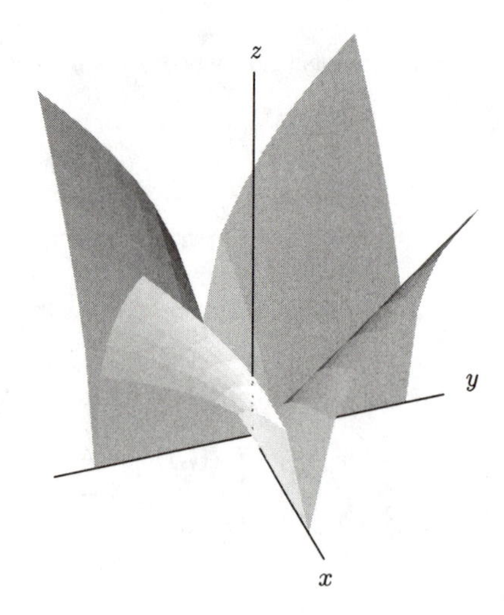

Figure 14.10

(c) f is clearly differentiable at (x, y) where $x \neq 0$ and $y \neq 0$. So we need to look at points $(x_0, 0)$, $x_0 \neq 0$ and $(0, y_0)$, $y_0 \neq 0$. At $(x_0, 0)$:

$$f_x(x_0, 0) = \lim_{x \to x_0} \frac{f(x, 0) - f(x_0, 0)}{x - x_0} = 0$$

$$f_y(x_0, 0) = \lim_{y \to 0} \frac{f(x_0, y) - f(x_0, 0)}{y} = \lim_{y \to 0} \frac{\sqrt{|x_0 y|}}{y}$$

which doesn't exist. So f is not differentiable at the points $(x_0, 0)$, $x_0 \neq 0$. Similarly, f is not differentiable at the points $(0, y_0)$, $y_0 \neq 0$.

(d)

$$f_x(0, 0) = \lim_{x \to 0} \frac{f(x, 0) - f(0, 0)}{x} = 0$$

$$f_y(0, 0) = \lim_{y \to 0} \frac{f(0, y) - f(0, 0)}{y} = 0$$

(e) Let $\vec{u} = (\vec{i} + \vec{j})/\sqrt{2}$:

$$f_{\vec{u}}(0, 0) = \lim_{t \to 0+} \frac{f(\frac{t}{\sqrt{2}}, \frac{t}{\sqrt{2}}) - f(0, 0)}{t} = \lim_{t \to 0+} \frac{\sqrt{\frac{t^2}{2}}}{t} = \frac{1}{\sqrt{2}}.$$

We know that $\nabla f(0, 0) = \vec{0}$ because both partial derivatives are 0. But if f were differentiable, $f_{\vec{u}}(0, 0) = \nabla f(0, 0) \cdot \vec{u} = f_x(0, 0) \cdot \frac{1}{\sqrt{2}} + f_y(0, 0) \cdot \frac{1}{\sqrt{2}} = 0$. But since, in fact, $f_{\vec{u}}(0, 0) = 1/\sqrt{2}$, we conclude that f is not differentiable.

Solutions for Chapter 14 Review

Exercises

1. $\dfrac{\partial z}{\partial x} = \dfrac{\partial}{\partial x}\left[(x^2 + x - y)^7\right] = 7(x^2 + x - y)^6(2x + 1) = (14x + 7)(x^2 + x - y)^6.$

$\dfrac{\partial z}{\partial y} = \dfrac{\partial}{\partial y}\left[(x^2 + x - y)^7\right] = -7(x^2 + x - y)^6.$

5. The first order partial derivative f_x is

$$f_x = -x(x^2 + y^2)^{-3/2}$$

Thus the second order partials are

$$f_{xx} = -(x^2 + y^2)^{-3/2} + 3x^2(x^2 + y^2)^{-5/2} = (2x^2 - y^2)(x^2 + y^2)^{-5/2}$$

and

$$f_{xy} = 3xy(x^2 + y^2)^{-5/2}.$$

9. Since the partial derivatives are

$$f_x = \frac{2x}{x^2+y^2} \quad \text{and} \quad f_y = \frac{2y}{x^2+y^2},$$

we have

$$\nabla f = \frac{2}{x^2+y^2}(x\vec{i} + y\vec{j}\,).$$

13. We have grad $f = 3x^2\vec{i} - 3y^2\vec{j}$, so grad $f(2,-1) = 12\vec{i} - 3\vec{j}$. A unit vector in the direction we want is $\vec{u} = (1/\sqrt{2})(\vec{i} - \vec{j}\,)$. Therefore, the directional derivative is

$$\text{grad}\, f(-2,1)\cdot\vec{u} = \frac{12\cdot 1 - 3(-1)}{\sqrt{2}} = \frac{15}{\sqrt{2}}.$$

17. The unit vector $\vec{u}$ in the direction of $\vec{v} = \vec{i} - \vec{k}$ is $\vec{u} = \frac{1}{\sqrt{2}}\vec{i} - \frac{1}{\sqrt{2}}\vec{k}$. We have

$$\begin{aligned} f_x(x,y,z) &= 6xy^2, && \text{and } f_x(-1,0,4) = 0 \\ f_y(x,y,z) &= 6x^2y + 2z, && \text{and } f_y(-1,0,4) = 8 \\ f_z(x,y,z) &= 2y, && \text{and } f_z(-1,0,4) = 0. \end{aligned}$$

So,

$$\begin{aligned} f_{\vec{u}}(-1,0,4) &= f_x(-1,0,4)\left(\frac{1}{\sqrt{2}}\right) + f_y(-1,0,4)(0) + f_z(-1,0,4)\left(-\frac{1}{\sqrt{2}}\right) \\ &= 0\left(\frac{1}{\sqrt{2}}\right) + 8(0) + 0\left(-\frac{1}{\sqrt{2}}\right) \\ &= 0. \end{aligned}$$

21. The given surface is the level surface for $f(x,y,z) = z^2 - 2xyz - x^2 - y^2$ passing through the point $(1,2,-1)$. Thus a normal vector is grad $f(1,2,-1)$. We have

$$\text{grad}\, f = (-2x - 2yz)\vec{i} + (-2y - 2xz)\vec{j} + (2z - 2xy)\vec{k},$$

so a normal vector is grad $f(1,2,-1) = 2\vec{i} - 2\vec{j} - 6\vec{k}$.

Problems

25. (a) The difference quotient for evaluating $f_w(2,2)$ is

$$\begin{aligned} f_w(2,2) &\approx \frac{f(2+0.01,2) - f(2,2)}{h} = \frac{e^{(2.01)\ln 2} - e^{2\ln 2}}{0.01} = \frac{e^{\ln(2^{2.01})} - e^{\ln(2^2)}}{0.01} \\ &= \frac{2^{(2.01)} - 2^2}{0.01} \approx 2.78 \end{aligned}$$

The difference quotient for evaluating $f_z(2,2)$ is

$$\begin{aligned} f_z(2,2) &\approx \frac{f(2,2+0.01) - f(2,2)}{h} \\ &= \frac{e^{2\ln(2.01)} - e^{2\ln 2}}{0.01} = \frac{(2.01)^2 - 2^2}{0.01} = 4.01 \end{aligned}$$

(b) Using the derivative formulas we get

$$\begin{aligned} f_w &= \frac{\partial f}{\partial w} = \ln z \cdot e^{w\ln z} = z^w \cdot \ln z \\ f_z &= \frac{\partial f}{\partial z} = e^{w\ln z}\cdot\frac{w}{z} = w\cdot z^{w-1} \end{aligned}$$

so

$$\begin{aligned} f_w(2,2) &= 2^2\cdot\ln 2 \approx 2.773 \\ f_z(2,2) &= 2\cdot 2^{2-1} = 4. \end{aligned}$$

29. Using local linearity with the slope in the x direction of -3 and slope in the y direction of 4, we get the values in Table 14.3.

Table 14.3

x \ y	0.9	1.0	1.1
1.8	7.2	7.6	8.0
2.0	6.6	7.0	7.4
2.2	6.0	6.4	6.8

33. To increase f as much as possible, we should head in the direction of the gradient from the point $(2,1)$. The rate of increase of f in the direction of the gradient is the magnitude of the gradient. Since at the point $(2,1)$, we know $\operatorname{grad} f = -3\vec{i} + 4\vec{j}$, the magnitude is 5. The furthest we can go from $(2,1)$, inside the circle, is 0.1 units, so the most we can increase f is $(0.1)(5) = 0.5$. Thus

$$\text{Largest value of function} \approx f(2,1) + 0.5 = 7.5.$$

This value is achieved at the point obtained from $(2,1)$ by a displacement of 0.1 units in the direction of $\operatorname{grad} f$, that is, a displacement by the vector

$$(0.1)\frac{\operatorname{grad} f}{\|\operatorname{grad} f\|} = (0.1)((-3/5)\vec{i} + (4/5)\vec{j}) = -0.06\vec{i} + 0.08\vec{j}.$$

Thus, the largest value of f is achieved at the point $(2-0.06, 1+0.08) = (1.94, 1.08)$.

37. The sign of $\partial f/\partial P_1$ tells you whether f (the number of people who ride the bus) increases or decreases when P_1 is increased. Since P_1 is the price of taking the bus, as it increases, f should decrease. This is because fewer people will be willing to pay the higher price, and more people will choose to ride the train. On the other hand, the sign of $\dfrac{\partial f}{\partial P_2}$ tells you the change in f as P_2 increases. Since P_2 is the cost of riding the train, as it increases, f should increase. This is because fewer people will be willing to pay the higher fares for the train, and more people will choose to ride the bus.

Therefore, $\dfrac{\partial f}{\partial P_1} < 0$ and $\dfrac{\partial f}{\partial P_2} > 0$.

41. The differential is

$$dP = \frac{\partial P}{\partial L}\,dL + \frac{\partial P}{\partial K}\,dK = 10L^{-0.75}K^{0.75}\,dL + 30L^{0.25}K^{-0.25}dK.$$

When $L = 2$ and $K = 16$, this is

$$dP \approx 47.6\,dL + 17.8\,dK.$$

45. $f_x = 2x$, $f_y = -2y$, so $\operatorname{grad} f(3,-1) = 6\vec{i} + 2\vec{j}$. For the direction $\theta = \pi/4$, the direction is $\vec{u} = \frac{1}{\sqrt{2}}\vec{i} + \frac{1}{\sqrt{2}}\vec{j}$, so $f_{\vec{u}}(3,-1) = (6\vec{i} + 2\vec{j}) \cdot (\frac{1}{\sqrt{2}}\vec{i} + \frac{1}{\sqrt{2}}\vec{j}) = \frac{8}{\sqrt{2}} = 4\sqrt{2}$.

The directional derivative is largest in the direction of the gradient vector $\operatorname{grad} f(3,-1) = 6\vec{i} + 2\vec{j}$.

49. **(a)** Fix $y = 3$. When x changes from 2.00 to 2.01, $f(x,3)$ decreases from 7.56 to 7.42. So

$$\left.\frac{\partial f}{\partial x}\right|_{(2,3)} \approx \left.\frac{\Delta f}{\Delta x}\right|_{(2,3)} = \frac{7.42 - 7.56}{2.01 - 2.00} = \frac{-0.14}{0.01} = -14.$$

Fix $x = 2$, when y changes from 3.00 to 3.02, $f(2,y)$ increases from 7.56 to 7.61. So

$$\left.\frac{\partial f}{\partial y}\right|_{(2,3)} \approx \left.\frac{\Delta f}{\Delta y}\right|_{(2,3)} = \frac{7.61 - 7.56}{3.02 - 3.00} = \frac{0.05}{0.02} = 2.5.$$

(b) Since the unit vector $\vec{u}$ of the direction $\vec{i} + 3\vec{j}$ is

$$\vec{u} = \frac{\vec{i} + 3\vec{j}}{\|\vec{i} + 3\vec{j}\|} = \frac{1}{\sqrt{10}}\vec{i} + \frac{3}{\sqrt{10}}\vec{j},$$

$$f_{\vec{u}}(2,3) = \text{grad}\, f(2,3)\cdot\vec{u} \approx \left(\frac{\Delta f}{\Delta x}\bigg|_{(2,3)} \vec{i} + \frac{\Delta f}{\Delta y}\bigg|_{(2,3)} \vec{j} \right)\cdot\vec{u}$$

$$= (-14\vec{i} + 2.5\vec{j})\cdot\left(\frac{1}{\sqrt{10}}\vec{i} + \frac{3}{\sqrt{10}}\vec{j}\right) = -\frac{6.5}{\sqrt{10}} \approx -2.055.$$

(c) Maximum rate equals $\|\text{grad}\, f\| \approx \sqrt{(-14)^2 + (2.5)^2} \approx 14.221$ in the direction of the gradient which is approximately equal to $-14\vec{i} + 2.5\vec{j}$.

(d) The equation of the level curve is

$$f(x,y) = f(2,3) = 7.56.$$

(e) The vector must be perpendicular to grad f, so $\vec{v} = 2.5\vec{i} + 14\vec{j}$ is a possible answer. (There are many others.).

(f) The differential at the point $(2,3)$ is

$$df = -14\, dx + 2.5\, dy.$$

If $dx = 0.03$, $dy = 0.04$, we get

$$df = -14(0.03) + 2.5(0.04) = -0.32.$$

The df approximates the change in f when (x,y) changes from $(2,3)$ to $(2.03, 3.04)$.

53. We know from Problem 52 that $V(x,y) = xF(2x+y)$ will solve the equation for any function F, so it is enough to find an F such that $xF(2x+y) = y^2$ when $x = 1$.
In other words, $1\cdot F(2+y) = y^2$, i.e. $F(y) = (y-2)^2$ for all y. So one solution is $V(x,y) = xF(2x+y) = x(2x+y-2)^2 = 4x^3 + xy^2 + 4x + 4x^2y - 8x^2 - 4xy$.

CAS Challenge Problems

57. **(a)** We have

$$\begin{aligned} f(1,2) &= A_0 + A_1 + 2A_2 + A_3 + 2A_4 + 4A_5, \\ f_x(1,2) &= A_1 + 2A_3 + 2A_4, \\ f_y(1,2) &= A_2 + A_4 + 4A_5, \\ f_{xx}(1,2) &= 2A_3, \\ f_{xy}(1,2) &= A_4, \\ f_{yy}(1,2) &= 2A_5. \end{aligned}$$

Thus $Q(x,y) = (A_0 + A_1 + 2A_2 + A_3 + 2A_4 + 4_A5) + (A_1 + 2A_3 + 2A_4)(x-1) + (A_2 + A_4 + 4A_5)(y-2) + A_3(x-1)^2 + A_4(x-1)(y-2) + A_5(y-2)^2$. Expanding this expression in powers of x and y, we find

$$Q(x,y) = A_0 + A_1x + A_2y + A_3x^2 + A_4xy + A_5y^2 = f(x,y)$$

.

(b) The quadratic expansion for f about $(1,2)$ is equal to f itself. This is because f is already a quadratic function, and is true about any point (a,b).

(c) The linear approximation of $f(x,y)$ at $(1,2)$ is:

$$A_0 + A_1 + 2A_2 + A_3 + 2A_4 + 4A_5 + (A_1 + 2A_3 + 2A_4)(x-1) + (A_2 + A_4 + 4A_5)(y-2).$$

When we expand this we get

$$A_0 - A_3 + A_4 - 4A_5 + (A_1 + 2A_3 + 2A_4)x + (A_2 + A_4 + 4A_5)y.$$

This is not the same as $f(x,y)$, and it is not even the same as the linear part of $f(x,y)$, namely $A_0 + A_1x + A_2y$.

CHECK YOUR UNDERSTANDING

1. True. This is the instantaneous rate of change of f in the x-direction at the point $(10,20)$.
5. True. The property $f_x(a,b) > 0$ means that f increases in the positive x-direction near (a,b), so f must decrease in the negative x-direction near (a,b).

9. True. A function with constant $f_x(x, y)$ and $f_y(x, y)$ has constant x-slope and constant y-slope, and therefore has a graph which is a plane.

13. True. Since g is a function of x only, it can be treated like a constant when taking the y partial derivative.

17. True. The definition of $f_x(x, y)$ is the limit of the difference quotient

$$f_x(x, y) = \lim_{h \to 0} \frac{f(x + h, y) - f(x, y)}{h}.$$

The symmetry of f gives

$$\frac{f(x + h, y) - f(x, y)}{h} = \frac{f(y, x + h) - f(y, x)}{h}.$$

The definition of $f_y(y, x)$ is the limit of the difference quotient

$$f_y(y, x) = \lim_{h \to 0} \frac{f(y, x + h) - f(y, x)}{h}.$$

Thus $f_x(x, y) = f_y(y, x)$.

21. False. The equation $z = 2 + 2x(x - 1) + 3y^2(y - 1)$ is not linear. The correct equation is $z = 2 + 2(x - 1) + 3(y - 1)$, which is obtained by evaluating the partial derivatives at the point $(1, 1)$.

25. False. The graph of f is a paraboloid, opening upward. The tangent plane to this surface at any point lies completely under the surface (except at the point of tangency). So the local linearization *underestimates* the value of f at nearby points.

29. True. If f is linear, then $f(x, y) = mx + ny + c$ for some m, n and c. So $f_x = m$ and $f_y = n$ giving $df = m\,dx + n\,dy$, which is linear in the variables dx and dy.

33. True. If $\vec{u}$ is a unit vector, then the directional derivative is given by the formula $f_{\vec{u}}(a, b) = \text{grad}\, f(a, b) \cdot \vec{u}$.

37. True. The gradient points in the direction of maximal increase of f, and the opposite direction gives the direction of maximum decrease for f.

41. True. It is the rate of change of f in the direction of $\vec{u}$ at the point (x_0, y_0).

45. Is never true. If $\|\text{grad}\, f\| = 0$, then $\text{grad}\, f = 0$, so $\text{grad}\, f \cdot \vec{u} = 0$ for any unit vector $\vec{u}$. Thus the directional derivative must be zero.

CHAPTER FIFTEEN

Solutions for Section 15.1

Exercises

1. The point A is not a critical point and the contour lines look like parallel lines. The point B is a critical point and is a local maximum; the point C is a saddle point.

5. The partial derivatives are $f_x(x,y) = 3x^2 - 3$ which vanishes for $x = \pm 1$ and $f_y(x,y) = 3y^2 - 3$ which vanishes for $y = \pm 1$. The points $(1,1)$, $(1,-1)$, $(-1,1)$, $(-1,-1)$ where both partials vanish are the critical points. To determine the nature of these critical points we calculate their discriminant and use the second derivative test. The discriminant is

$$D = f_{xx}(x,y)f_{yy}(x,y) - f_{xy}^2(x,y) = (6x)(6y) - 0 = 36xy.$$

At $(1,1)$ and $(-1,-1)$ the discriminant is positive. Since $f_{xx}(1,1) = 6$ is positive, $(1,1)$ is a local minimum. And since $f_{xx}(-1,-1) = -6$ is negative, $(-1,-1)$ is a local maximum. The remaining two points, $(1,-1)$ and $(-1,1)$ are saddle points since the discriminant is negative.

9. To find the critical points, we solve $f_x = 0$ and $f_y = 0$ for x and y. Solving

$$f_x = 3x^2 - 3 = 0$$
$$f_y = 3y^2 - 12y = 0$$

shows that $x = -1$ or $x = 1$ and $y = 0$ or $y = 4$. There are four critical points: $(-1,0)$, $(1,0)$, $(-1,4)$, and $(1,4)$.

We have
$$D = (f_{xx})(f_{yy}) - (f_{xy})^2 = (6x)(6y-12) - (0)^2 = (6x)(6y-12).$$

At critical point $(-1,0)$, we have $D > 0$ and $f_{xx} < 0$, so f has a local maximum at $(-1,0)$.
At critical point $(1,0)$, we have $D < 0$, so f has a saddle point at $(1,0)$.
At critical point $(-1,4)$, we have $D < 0$, so f has a saddle point at $(-1,4)$.
At critical point $(1,4)$, we have $D > 0$ and $f_{xx} > 0$, so f has a local minimum at $(1,4)$.

13. At a critical point, $f_x = 0$, $f_y = 0$.

$$f_x = 8y - (x+y)^3 = 0, \text{ we know } 8y = (x+y)^3.$$

$$f_y = 8x - (x+y)^3 = 0, \text{ we know } 8x = (x+y)^3.$$

Therefore we must have $x = y$. Since $(x+y)^3 = (2y)^3 = 8y^3$, this tells us that $8y - 8y^3 = 0$. Solving gives $y = 0, \pm 1$. Thus the critical points are $(0,0)$, $(1,1)$, $(-1,-1)$.

$f_{yy} = f_{xx} = -3(x+y)^2$, and $f_{xy} = 8 - 3(x+y)^2$.
The discriminant is

$$\begin{aligned} D(x,y) &= f_{xx}f_{yy} - f_{xy}^2 \\ &= 9(x+y)^4 - \left(64 - 48(x+y)^2 + 9(x+y)^4\right) \\ &= -64 + 48(x+y)^2. \end{aligned}$$

$D(0,0) = -64 < 0$, so $(0,0)$ is a saddle point.
$D(1,1) = -64 + 192 > 0$ and $f_{xx}(1,1) = -12 < 0$, so $(1,1)$ is a local maximum.
$D(-1,-1) = -64 + 192 > 0$ and $f_{xx}(-1,-1) = -12 < 0$, so $(-1,-1)$ is a local maximum.

Problems

17. To find critical points, set partial derivatives equal to zero:

$$E_x = \sin x = 0 \quad \text{when} \quad x = 0, \pm\pi, \pm 2\pi, \cdots$$

$$E_y = y = 0 \quad \text{when} \quad y = 0.$$

The critical points are

$$\cdots(-2\pi, 0), (-\pi, 0), (0, 0), (\pi, 0), (2\pi, 0), (3\pi, 0)\cdots$$

To classify, calculate $D = E_{xx}E_{yy} - (E_{xy})^2 = \cos x$.
At the points $(0, 0), (\pm 2\pi, 0), (\pm 4\pi, 0), (\pm 6\pi, 0), \cdots$

$$D = (1) > 0 \quad \text{and} \quad E_{xx} > 0 \quad (\text{Since} E_{xx}(0, 2k\pi) = \cos(2k\pi) = 1).$$

Therefore $(0, 0), (\pm 2\pi, 0), (\pm 4\pi, 0), (\pm 6\pi, 0), \cdots$ are local minima.
At the points $(\pm\pi, 0), (\pm 3\pi, 0), (\pm 5\pi, 0), (\pm 7\pi, 0), \cdots$, we have $\cos(2k+1)\pi = -1$, so

$$D = (-1) < 0.$$

Therefore $(\pm\pi, 0), (\pm 3\pi, 0), (\pm 5\pi, 0), (\pm 7\pi, 0), \cdots$ are saddle points.

21. **(a)** $(1, 3)$ is a critical point. Since $f_{xx} > 0$ and the discriminant

$$D = f_{xx}f_{yy} - f_{xy}^2 = f_{xx}f_{yy} - 0^2 = f_{xx}f_{yy} > 0,$$

the point $(1, 3)$ is a minimum.

(b)

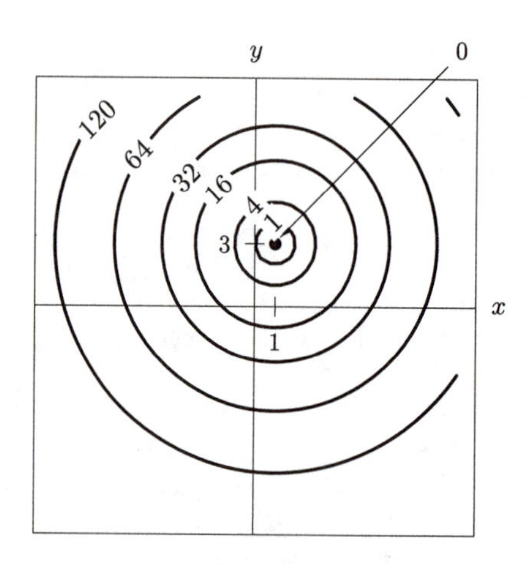

Figure 15.1

25. The first order partial derivatives are

$$f_x(x, y) = 2kx - 2y \qquad \text{and} \qquad f_y(x, y) = 2ky - 2x.$$

And the second order partial derivatives are

$$f_{xx}(x, y) = 2k \qquad f_{xy}(x, y) = -2 \qquad f_{yy}(x, y) = 2k$$

Since $f_x(0, 0) = f_y(0, 0) = 0$, the point $(0, 0)$ is a critical point. The discriminant is

$$D = (2k)(2k) - 4 = 4(k^2 - 1).$$

For $k = \pm 2$, the discriminant is positive, $D = 12$. When $k = 2$, $f_{xx}(0, 0) = 4$ which is positive so we have a local minimum at the origin. When $k = -2$, $f_{xx}(0, 0) = -4$ so we have a local maximum at the origin. In the case $k = 0$, $D = -4$ so the origin is a saddle point.

Lastly, when $k = \pm 1$ the discriminant is zero, so the second derivative test can tell us nothing. Luckily, we can factor $f(x, y)$ when $k = \pm 1$. When $k = 1$,

$$f(x, y) = x^2 - 2xy + y^2 = (x - y)^2.$$

This is always greater than or equal to zero. So $f(0, 0) = 0$ is a minimum and the surface is a trough-shaped parabolic cylinder with its base along the line $x = y$.

When $k = -1$,

$$f(x, y) = -x^2 - 2xy - y^2 = -(x + y)^2.$$

This is always less than or equal to zero. So $f(0,0) = 0$ is a maximum. The surface is a parabolic cylinder, with its top ridge along the line $x = -y$.

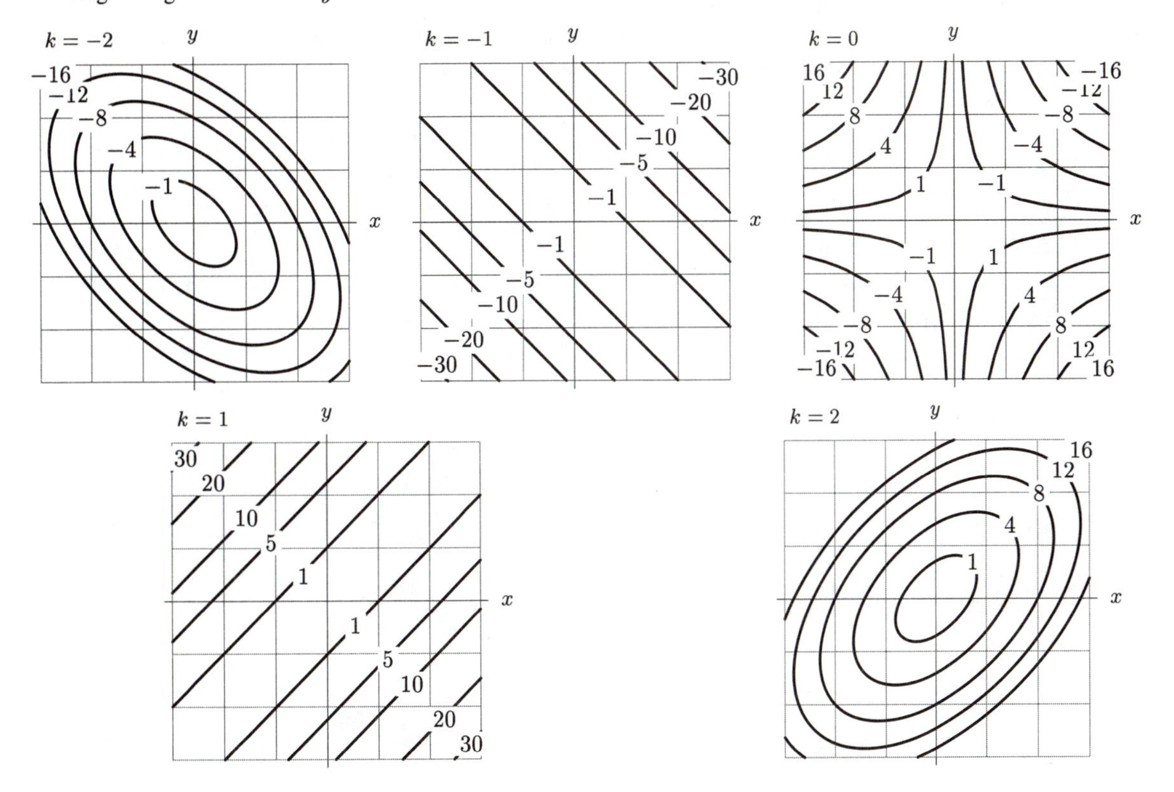

Figure 15.2

Solutions for Section 15.2

Exercises

1. Mississippi lies entirely within a region designated as 80s so we expect both the maximum and minimum daily high temperatures within the state to be in the 80s. The southwestern-most corner of the state is close to a region designated as 90s, so we would expect the temperature here to be in the high 80s, say 87-88. The northern-most portion of the state is located near the center of the 80s region. We might expect the high temperature there to be between 83-87.

Alabama also lies completely within a region designated as 80s so both the high and low daily high temperatures within the state are in the 80s. The southeastern tip of the state is close to a 90s region so we would expect the temperature here to be about 88-89 degrees. The northern-most part of the state is near the center of the 80s region so the temperature there is 83-87 degrees.

Pennsylvania is also in the 80s region, but it is touched by the boundary line between the 80s and a 70s region. Thus we expect the low daily high temperature to occur there and be about 80 degrees. The state is also touched by a boundary line of a 90s region so the high will occur there and be 89-90 degrees.

New York is split by a boundary between an 80s and a 70s region, so the northern portion of the state is likely to be about 74-76 while the southern portion is likely to be in the low 80s, maybe 81-84 or so.

California contains many different zones. The northern coastal areas will probably have the daily high as low as 65-68, although without another contour on that side, it is difficult to judge how quickly the temperature is dropping off to the west. The tip of Southern California is in a 100s region, so there we expect the daily high to be 100-101.

Arizona will have a low daily high around 85-87 in the northwest corner and a high in the 100s, perhaps 102-107 in its southern regions.

Massachusetts will probably have a high daily high around 81-84 and a low daily high of 70.

5. To maximize $z = x^2 + y^2$, it suffices to maximize x^2 and y^2. We can maximize both of these at the same time by taking the point $(1, 1)$, where $z = 2$. It occurs on the boundary of the square. (Note: We also have maxima at the points $(-1, -1)$, $(-1, 1)$ and $(1, -1)$ which are on the boundary of the square.)

To minimize $z = x^2 + y^2$, we choose the point $(0, 0)$, where $z = 0$. It does not occur on the boundary of the square.

9. The maximum value, which is slightly above 30, say 30.5, occurs approximately at the origin. The minimum value, which is about 20.5, occurs at $(2.5, 5)$.

Problems

13. Let the sides be x, y, z cm. Then the volume is given by $V = xyz = 32$.

The surface area S is given by

$$S = 2xy + 2xz + 2yz.$$

Substituting $z = 32/(xy)$ gives

$$S = 2xy + \frac{64}{y} + \frac{64}{x}.$$

At a critical point,

$$\frac{\partial S}{\partial x} = 2y - \frac{64}{x^2} = 0$$
$$\frac{\partial S}{\partial y} = 2x - \frac{64}{y^2} = 0,$$

The symmetry of the equations (or by dividing the equations) tells us that $x = y$ and

$$2x - \frac{64}{x^2} = 0$$
$$x^3 = 32$$
$$x = 32^{1/3} = 3.17 \text{ cm}.$$

Thus the only critical point is $x = y = (32)^{1/3}$ cm and $z = 32/\left((32)^{1/3} \cdot (32)^{1/3}\right) = (32)^{1/3}$ cm. At the critical point

$$S_{xx}S_{yy} - (S_{xy})^2 = \frac{128}{x^3} \cdot \frac{128}{y^3} - 2^2 = \frac{(128)^2}{x^3y^3} - 4.$$

Since $D > 0$ and $S_{xx} > 0$ at this critical point, the critical point $x = y = z = (32)^{1/3}$ is a local minimum. Since $S \to \infty$ as $x, y \to \infty$, the local minimum is a global minimum.

17. We minimize the square of the distance from the point (x, y, z) to the origin:

$$S = x^2 + y^2 + z^2.$$

Since $z^2 = 9 - xy - 3x$, we have

$$S = x^2 + y^2 + 9 - xy - 3x.$$

At a critical point

$$\frac{\partial S}{\partial x} = 2x - y - 3 = 0$$
$$\frac{\partial S}{\partial y} = 2y - x = 0,$$

so $x = 2y$, and

$$2(2y) - y - 3 = 0$$

giving $y = 1$, so $x = 2$ and $z^2 = 9 - 2 \cdot 1 - 3 \cdot 2 = 1$, so $z = \pm 1$. We have

$$D = S_{xx}S_{yy} - (S_{xy})^2 = 2 \cdot 2 - (-1)^2 = 4 - 1 > 0,$$

so, since $D > 0$ and $S_{xx} > 0$, the critical points are local minima. Since $S \to \infty$ as $x, y \to \pm\infty$, the local minima are global minima.

If $x = 2, y = 1, z = \pm 1$, we have $S = 2^2 + 1^2 + 1^2 = 6$, so the shortest distance to the origin is $\sqrt{6}$.

21. We calculate the partial derivatives and set them to zero.

$$\frac{\partial\,(\text{range})}{\partial t} = -10t - 6h + 400 = 0$$
$$\frac{\partial\,(\text{range})}{\partial h} = -6t - 6h + 300 = 0.$$

$$10t + 6h = 400$$
$$6t + 6h = 300$$

solving we obtain

$$4t = 100$$

so

$$t = 25$$

Solving for h, we obtain $6h = 150$, yielding $h = 25$. Since the range is quadratic in h and t, the second derivative test tells us this is a local and global maximum. So the optimal conditions are $h = 25\%$ humidity and $t = 25^\circ$C.

25. Let $P(K, L)$ be the profit obtained using K units of capital and L units of labor. The cost of production is given by

$$C(K, L) = kK + \ell L,$$

and the revenue function is given by

$$R(K, L) = pQ = pAK^aL^b.$$

Hence, the profit is

$$P = R - C = pAK^aL^b - (kK + \ell L).$$

In order to find local maxima of P, we calculate the partial derivatives and see where they are zero. We have:

$$\frac{\partial P}{\partial K} = apAK^{a-1}L^b - k,$$
$$\frac{\partial P}{\partial L} = bpAK^aL^{b-1} - \ell.$$

The critical points of the function $P(K, L)$ are solutions (K, L) of the simultaneous equations:

$$\frac{k}{a} = pAK^{a-1}L^b,$$
$$\frac{\ell}{b} = pAK^aL^{b-1}.$$

Multiplying the first equation by K and the second by L, we get

$$\frac{kK}{a} = \frac{\ell L}{b},$$

and so

$$K = \frac{\ell a}{kb}L.$$

Substituting for K in the equation $k/a = pAK^{a-1}L^b$, we get:

$$\frac{k}{a} = pA\left(\frac{\ell a}{kb}\right)^{a-1} L^{a-1}L^b.$$

We must therefore have

$$L^{1-a-b} = pA\left(\frac{a}{k}\right)^a\left(\frac{\ell}{b}\right)^{a-1}.$$

Hence, if $a + b \neq 1$,

$$L = \left[pA\left(\frac{a}{k}\right)^a\left(\frac{\ell}{b}\right)^{(a-1)}\right]^{1/(1-a-b)},$$

and

$$K = \frac{\ell a}{kb}L = \frac{\ell a}{kb}\left[pA\left(\frac{a}{k}\right)^a\left(\frac{\ell}{b}\right)^{(a-1)}\right]^{1/(1-a-b)}.$$

To see if this is really a local maximum, we apply the second derivative test. We have:

$$\frac{\partial^2 P}{\partial K^2} = a(a-1)pAK^{a-2}L^b,$$
$$\frac{\partial^2 P}{\partial L^2} = b(b-1)pAK^a L^{b-2},$$
$$\frac{\partial^2 P}{\partial K \partial L} = abpAK^{a-1}L^{b-1}.$$

Hence,

$$\begin{aligned} D &= \frac{\partial^2 P}{\partial K^2}\frac{\partial^2 P}{\partial L^2} - \left(\frac{\partial^2 P}{\partial K \partial L}\right)^2 \\ &= ab(a-1)(b-1)p^2A^2K^{2a-2}L^{2b-2} - a^2b^2p^2A^2K^{2a-2}L^{2b-2} \\ &= ab((a-1)(b-1) - ab)p^2A^2K^{2a-2}L^{2b-2} \\ &= ab(1-a-b)p^2A^2K^{2a-2}L^{2b-2}. \end{aligned}$$

Now $a, b, p, A, K,$ and L are positive numbers. So, the sign of this last expression is determined by the sign of $1-a-b$.

(a) We assumed that $a+b<1$, so $D>0$, and as $0<a<1$, then $\partial^2 P/\partial K^2 < 0$ and so we have a unique local maximum. To verify that the local maximum is a global maximum, we focus on the cost. Let $C = kK + \ell L$. Since $K \geq 0$ and $L \geq 0$, $K \leq C/k$ and $L \leq C/\ell$. Therefore the profit satisfies:

$$\begin{aligned} P &= pAK^aL^b - (kK + \ell L) \\ &\leq pA\left(\frac{C}{k}\right)^a\left(\frac{C}{\ell}\right)^b - C \\ &= mC^{a+b} - C \end{aligned}$$

where $m = pA(1/k)^a(1/\ell)^b$. Since $a+b<1$, the profit is negative for large costs C, say $C \geq C_0$ ($C_0 = m^{1-a-b}$ will do). Therefore, in the KL-plane for $K \geq 0$ and $L \geq 0$, the profit is less than or equal to zero everywhere on or above the line $kK + \ell L = C_o$. Thus the global maximum must occur inside the triangle bounded by this line and the K and L axes. Since $P \leq 0$ on the K and L axes as well, the global maximum must be in the interior of the triangle at the unique local maximum we found.

In the case $a+b<1$, we have decreasing returns to scale. That is, if the amount of capital and labor used is multiplied by a constant $\lambda > 0$, we get less than λ times the production.

(b) Now suppose $a+b \geq 1$. If we multiply K and L by λ for some $\lambda > 0$, then

$$Q(\lambda K, \lambda L) = A(\lambda K)^a(\lambda L)^b = \lambda^{a+b}Q(K,L).$$

We also see that

$$C(\lambda K, \lambda L) = \lambda C(K,L).$$

So if $a+b=1$, we have

$$P(\lambda K, \lambda L) = \lambda P(K,L).$$

Thus, if $\lambda = 2$, so we are doubling the inputs K and L, then the profit P is doubled and hence there can be no maximum profit.

If $a+b>1$, we have increasing returns to scale and there can again be no maximum profit: doubling the inputs will more than double the profit. In this case, the profit increases without bound as $K,\ L$ go toward infinity.

29. **(a)** The function f is continuous in the region R, but R is not closed and bounded so a special analysis is required.

Notice that $f(x,y)$ tends to ∞ as (x,y) tends farther and farther from the origin or tends toward any point on the x or y axis. This suggests that a minimum for f, if it exists, can not be too far from the origin or too close to the axes. For example, if $x>10$ then $f(x,y) > 4x > 40$, and if $y > 10$ then $f(x,y) > 5y > 50$. If $0 < x < 0.1$ then $f(x,y) > 2/x > 20$, and if $0 < y < 0.1$ then $f(x,y) > 3/y > 30$.

Since $f(1,1) = 14$, a global minimum for f if it exists must be in the smaller region $R' : 0.1 \leq x \leq 10$, $0.1 \leq y \leq 10$. The region R' is closed and bounded and so f does have a minimum value at some point in R', and since that value is at most 14, it is also a global minimum for all of R.

(b) Since the region R has no boundary, the minimum value must occur at a critical point of f. At a critical point we have

$$f_x = -\frac{2}{x^2} + 4 = 0 \qquad f_y = -\frac{3}{y^2} + 5 = 0.$$

The only critical point is $(\sqrt{1/2}, \sqrt{3/5}) \approx (0.7071, 0.7746)$, at which f achieves the minimum value $f(\sqrt{1/2}, \sqrt{3/5}) = 4\sqrt{2} + 2\sqrt{15} \approx 13.403$.

Solutions for Section 15.3

Exercises

1. Our objective function is $f(x,y) = x + y$ and our equation of constraint is $g(x,y) = x^2 + y^2 = 1$. To optimize $f(x,y)$ with Lagrange multipliers, we solve $\nabla f(x,y) = \lambda \nabla g(x,y)$ subject to $g(x,y) = 1$. The gradients of f and g are

$$\nabla f(x,y) = \vec{i} + \vec{j},$$
$$\nabla g(x,y) = 2x\vec{i} + 2y\vec{j}.$$

So the equation $\nabla f = \lambda \nabla g$ becomes

$$\vec{i} + \vec{j} = \lambda(2x\vec{i} + 2y\vec{j})$$

Solving for λ gives

$$\lambda = \frac{1}{2x} = \frac{1}{2y},$$

which tells us that $x = y$. Going back to our equation of constraint, we use the substitution $x = y$ to solve for y:

$$g(y,y) = y^2 + y^2 = 1$$
$$2y^2 = 1$$
$$y^2 = \frac{1}{2}$$
$$y = \pm\sqrt{\frac{1}{2}} = \pm\frac{\sqrt{2}}{2}.$$

Since $x = y$, our critical points are $(\frac{\sqrt{2}}{2}, \frac{\sqrt{2}}{2})$ and $(-\frac{\sqrt{2}}{2}, -\frac{\sqrt{2}}{2})$. Since the constraint is closed and bounded, maximum and minimum values of f subject to the constraint exist. Evaluating f at the critical points we find that the maximum value is $f(\frac{\sqrt{2}}{2}, \frac{\sqrt{2}}{2}) = \sqrt{2}$ and the minimum value is $f(-\frac{\sqrt{2}}{2}, -\frac{\sqrt{2}}{2}) = -\sqrt{2}$.

5. Our objective function is $f(x,y) = xy$ and our equation of constraint is $g(x,y) = 4x^2 + y^2 = 8$. Their gradients are

$$\nabla f(x,y) = y\vec{i} + x\vec{j},$$
$$\nabla g(x,y) = 8x\vec{i} + 2y\vec{j}.$$

So the equation $\nabla f = \lambda \nabla g$ becomes $y\vec{i} + x\vec{j} = \lambda(8x\vec{i} + 2y\vec{j})$. This gives

$$8x\lambda = y \quad \text{and} \quad 2y\lambda = x.$$

Multiplying, we get

$$8x^2\lambda = 2y^2\lambda.$$

If $\lambda = 0$, then $x = y = 0$, which doesn't satisfy the constraint equation. So $\lambda \neq 0$ and we get

$$2y^2 = 8x^2$$
$$y^2 = 4x^2$$
$$y = \pm 2x.$$

To find x, we substitute for y in our equation of constraint.

$$4x^2 + y^2 = 8$$
$$4x^2 + 4x^2 = 8$$
$$x^2 = 1$$
$$x = \pm 1$$

So our critical points are $(1,2)$, $(1,-2)$, $(-1,2)$ and $(-1,-2)$. Since the constraint is closed and bounded, maximum and minimum values of f subject to the constraint exist. Evaluating $f(x,y)$ at the critical points, we have

$$f(1,2) = f(-1,-2) = 2$$
$$f(1,-2) = f(1,-2) = -2.$$

So the maximum value of f on $g(x,y) = 8$ is 2, and the minimum value is -2.

9. Our objective function is $f(x, y, z) = 2x + y + 4z$ and our equation of constraint is $g(x, y, z) = x^2 + y + z^2 = 16$. Their gradients are

$$\nabla f(x, y, z) = 2\vec{i} + 1\vec{j} + 4\vec{k},$$
$$\nabla g(x, y, z) = 2x\vec{i} + 1\vec{j} + 2z\vec{k}.$$

So the equation $\nabla f = \lambda \nabla g$ becomes $2\vec{i} + 1\vec{j} + 4\vec{k} = \lambda(2x\vec{i} + 1\vec{j} + 2z\vec{k})$. Solving for λ we find

$$\lambda = \frac{2}{2x} = \frac{1}{1} = \frac{4}{2z}$$
$$\lambda = \frac{1}{x} = 1 = \frac{2}{z}.$$

Which tells us that $x = 1$ and $z = 2$. Going back to our equation of constraint, we can solve for y.

$$g(1, y, 2) = 16$$
$$1^2 + y + 2^2 = 16$$
$$y = 11.$$

So our one critical point is at $(1, 11, 2)$. The value of f at this point is $f(1, 11, 2) = 2 + 11 + 8 = 21$. This is the maximum value of $f(x, y, z)$ on $g(x, y, z) = 16$. To see this, note that for $y = 16 - x^2 - z^2$,

$$f(x, y, z) = 2x + 16 - x^2 - z^2 + 4z = 21 - (x - 1)^2 - (z - 2)^2 \le 21.$$

As $y \to -\infty$, the point $(-\sqrt{16 - y}, y, 0)$ is on the constraint and $f(-\sqrt{16 - y}, y, 0) \to -\infty$, so there is no minimum value for $f(x, y, z)$ on $g(x, y, z) = 16$.

13. The region $x^2 + y^2 \le 2$ is the shaded disk of radius $\sqrt{2}$ centered at the origin (including the circle $x^2 + y^2 = 2$) as shown in Figure 15.3.

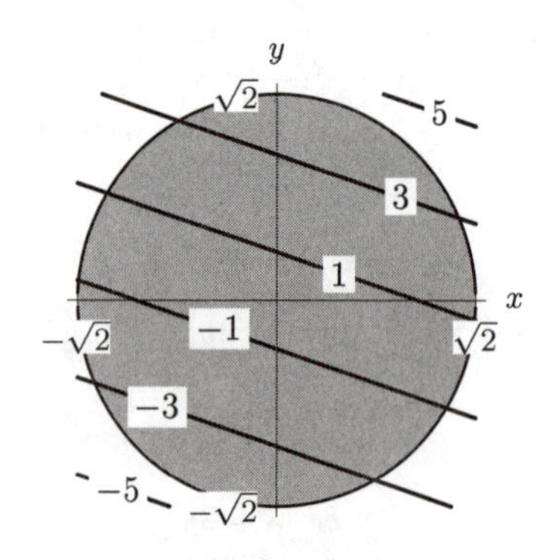

Figure 15.3

We first find the local maxima and minima of f in the interior of our disk. So we need to find the extrema of

$$f(x, y) = x + 3y, \quad \text{in the region} \quad x^2 + y^2 < 2.$$

As

$$f_x = 1$$
$$f_y = 3$$

f doesn't have critical points. Now let's find the local extrema of f on the boundary of the disk. We want to find the extrema of $f(x, y) = x + 3y$ subject to the constraint $g(x, y) = x^2 + y^2 - 2 = 0$. We use Lagrange multipliers

$$\text{grad}\, f = \lambda\, \text{grad}\, g \quad \text{and} \quad x^2 + y^2 = 2,$$

which give

$$1 = 2\lambda x$$
$$3 = 2\lambda y$$
$$x^2 + y^2 = 2.$$

As λ cannot be zero, we solve for x and y in the first two equations and get $x = \frac{1}{2\lambda}$ and $y = \frac{3}{2\lambda}$. Plugging into the third equation gives

$$8\lambda^2 = 10$$

so $\lambda = \pm\frac{\sqrt{5}}{2}$ and we get the solutions $(\frac{1}{\sqrt{5}}, \frac{3}{\sqrt{5}})$ and $(-\frac{1}{\sqrt{5}}, -\frac{3}{\sqrt{5}})$. Evaluating f at these points gives

$$f(\frac{1}{\sqrt{5}}, \frac{3}{\sqrt{5}}) = 2\sqrt{5} \quad \text{and}$$
$$f(-\frac{1}{\sqrt{5}}, -\frac{3}{\sqrt{5}}) = -2\sqrt{5}$$

The region $x^2 + y^2 \leq 2$ is closed and bounded, so maximum and minimum values of f in the region exist. Therefore $(\frac{1}{\sqrt{5}}, \frac{3}{\sqrt{5}})$ is a global maximum of f and $(-\frac{1}{\sqrt{5}}, -\frac{3}{\sqrt{5}})$ is a global minimum of f on the whole region $x^2 + y^2 \leq 2$.

Problems

17. **(a)** The contour for $z = 1$ is the line $1 = 2x + y$, or $y = -2x + 1$. The contour for $z = 3$ is the line $3 = 2x + y$, or $y = -2x + 3$. The contours are all lines with slope -2. See Figure 15.4.

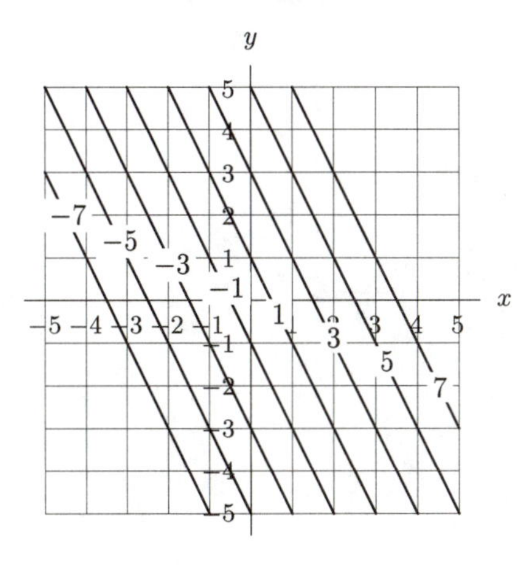

Figure 15.4

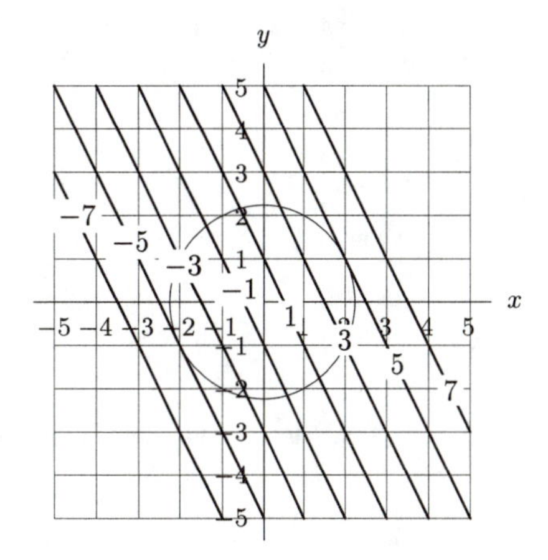

Figure 15.5

(b) The graph of $x^2 + y^2 = 5$ is a circle of radius $\sqrt{5} = 2.236$ centered at the origin. See Figure 15.5.

(c) The circle representing the constraint equation in Figure 15.5 appears to be tangent to the contour close to $z = 5$ at the point $(2, 1)$, and this is the contour with the highest z-value that the circle intersects. The circle is tangent to the contour $z = -5$ approximately at the point $(-2, -1)$, and this is the contour with the lowest z-value that the circle intersects. Therefore, subject to the constraint $x^2 + y^2 = 5$, the function f has a maximum value of about 5 at the point $(2, 1)$ and a minimum value of about -5 at the point $(-2, -1)$.

Since the radius vector, $2\vec{i} + \vec{j}$, at the point $(2, 1)$ is perpendicular to the line $2x + y = 5$, the maximum is exactly 5 and occurs at $(2, 1)$. Similarly, the minimum is exactly -5 and occurs at $(-2, -1)$.

(d) The objective function is $f(x, y) = 2x + y$ and the constraint equation is $g(x, y) = x^2 + y^2 = 5$, and so grad $f = 2\vec{i} + \vec{j}$ and grad $g = (2x)\vec{i} + (2y)\vec{j}$. Setting grad $f = \lambda$ grad g gives

$$2 = \lambda(2x),$$
$$1 = \lambda(2y).$$

On the constraint, $x \neq 0$ and $y \neq 0$. Thus, from the first equation, we have $\lambda = 1/x$, and from the second equation we have $\lambda = 1/(2y)$. Setting these equal gives

$$x = 2y.$$

Substituting this into the constraint equation $x^2 + y^2 = 5$ gives $(2y)^2 + y^2 = 5$ so $y = -1$ and $y = 1$. Since $x = 2y$, the maximum or minimum values occur at $(2, 1)$ or $(-2, -1)$. Since $f(2, 1) = 5$ and $f(-2, -1) = -5$, the function $f(x, y) = 2x + y$ subject to the constraint $x^2 + y^2 = 5$ has a maximum value of 5 at the point $(2, 1)$ and a minimum value of -5 at the point $(-2, -1)$. This confirms algebraically what we observed graphically in part (c).

21. **(a)** The problem is to maximize

$$V = 1000D^{0.6}N^{0.3}$$

subject to the budget constraint in dollars

$$40000D + 10000N \leq 600000$$

or (in thousand dollars)

$$40D + 10N \leq 600$$

(b) Let $B = 40D + 10N = 600$ (thousand dollars) be the budget constraint. At the optimum

$$\nabla V = \lambda \nabla B,$$

$$\text{so} \quad \frac{\partial V}{\partial D} = \lambda \frac{\partial B}{\partial D} = 40\lambda$$

$$\frac{\partial V}{\partial N} = \lambda \frac{\partial B}{\partial N} = 10\lambda.$$

$$\text{Thus} \quad \frac{\frac{\partial V}{\partial D}}{\frac{\partial V}{\partial N}} = 4.$$

Therefore, at the optimum point, the rate of increase in the number of visits to the number of doctors is four times the corresponding rate for nurses. This factor of four is the same as the ratio of the salaries.

(c) Differentiating and setting $\nabla V = \lambda \nabla B$ yields

$$600D^{-0.4}N^{0.3} = 40\lambda$$

$$300D^{0.6}N^{-0.7} = 10\lambda$$

Thus, we get

$$\frac{600D^{-0.4}N^{0.3}}{40} = \lambda = \frac{300D^{0.6}N^{-0.7}}{10}$$

So

$$N = 2D.$$

To solve for D and N, substitute in the budget constraint:

$$600 - 40D - 10N = 0$$

$$600 - 40D - 10 \cdot (2D) = 0$$

So $D = 10$ and $N = 20$.

$$\lambda = \frac{600(10^{-0.4})(20^{0.3})}{40} \approx 14.67$$

Thus the clinic should hire 10 doctors and 20 nurses. With that staff, the clinic can provide

$$V = 1000(10^{0.6})(20^{0.3}) \approx 9{,}779 \text{ visits per year.}$$

(d) From part c), the Lagrange multiplier is $\lambda = 14.67$. At the optimum, the Lagrange multiplier tells us that about 14.67 extra visits can be generated through an increase of \$1,000 in the budget. (If we had written out the constraint in dollars instead of thousands of dollars, the Lagrange multiplier would tell us the number of extra visits per dollar.)

(e) The marginal cost, MC, is the cost of an additional visit. Thus, at the optimum point, we need the reciprocal of the Lagrange multiplier:

$$\text{MC} = \frac{1}{\lambda} \approx \frac{1}{14.67} \approx 0.068 \ \text{(thousand dollars)}$$

i.e. at the optimum point, an extra visit costs the clinic 0.068 thousand dollars, or \$68.

This production function exhibits declining returns to scale (e.g. doubling both inputs less than doubles output, because the two exponents add up to less than one). This means that for large V, increasing V will require increasing D and N by more than when V is small. Thus the cost of an additional visit is greater for large V than for small. In other words, the marginal cost will rise with the number of visits.

25. Constraint is $G = P_1x + P_2y - K = 0$.
Since $\nabla Q = \lambda \nabla G$, we have

$$cax^{a-1}y^b = \lambda P_1 \quad \text{and} \quad cbx^a y^{b-1} = \lambda P_2.$$

Dividing the two equations yields $\dfrac{cax^{a-1}y^b}{cbx^a y^{b-1}} = \dfrac{\lambda P_1}{\lambda P_2}$, or simplifying, $\dfrac{ay}{bx} = \dfrac{P_1}{P_2}$. Hence, $y = \dfrac{bP_1}{aP_2}x$.

Substitute into the constraint to obtain $P_1x + P_2\dfrac{bP_1}{aP_2}x = P_1\left(\dfrac{a+b}{a}\right)x = K$, giving

$$x = \frac{aK}{(a+b)P_1} \quad \text{and} \quad y = \frac{bK}{(a+b)P_2}.$$

We now check that this is indeed the maximization point. Since $x, y \geq 0$, possible maximization points are $(0, \dfrac{K}{P_2})$, $(\dfrac{K}{P_1}, 0)$, and $(\dfrac{aK}{(a+b)P_1}, \dfrac{bK}{(a+b)P_2})$. Since $Q = 0$ for the first two points and Q is positive for the last point, it follows that $(\dfrac{aK}{(a+b)P_1}, \dfrac{bK}{(a+b)P_2})$ gives the maximal value.

29. We want to minimize the function $h(x, y)$ subject to the constraint that

$$g(x, y) = x^2 + y^2 = 1{,}000^2 = 1{,}000{,}000.$$

Using the method of Lagrange multipliers, we obtain the following system of equations:

$$\begin{aligned} h_x &= -\frac{10x + 4y}{10{,}000} = 2\lambda x, \\ h_y &= -\frac{4x + 4y}{10{,}000} = 2\lambda y, \\ x^2 + y^2 &= 1{,}000{,}000. \end{aligned}$$

Multiplying the first equation by y and the second by x we get

$$\frac{-y(10x + 4y)}{10{,}000} = \frac{-x(4x + 4y)}{10{,}000}.$$

Hence:

$$2y^2 + 3xy - 2x^2 = (2y - x)(y + 2x) = 0,$$

and so the climber either moves along the line $x = 2y$ or $y = -2x$.

We must now choose one of these lines and the direction along that line which will lead to the point of minimum height on the circle. To do this we find the points of intersection of these lines with the circle $x^2 + y^2 = 1{,}000{,}000$, compute the corresponding heights, and then select the minimum point.

If $x = 2y$, the third equation gives

$$5y^2 = 1{,}000^2,$$

so that $y = \pm 1{,}000/\sqrt{5} \approx \pm 447.21$ and $x = \pm 894.43$. The corresponding height is $h(\pm 894.43, \pm 447.21) = 2400$ m. If $y = -2x$, we find that $x = \pm 447.21$ and $y = \mp 894.43$. The corresponding height is $h(\pm 447.21, \mp 894.43) = 2900$ m. Therefore, she should travel along the line $x = 2y$, in either of the two possible directions.

33. **(a)** The objective function $f(x, y) = px + qy$ gives the cost to buy x units of input 1 at unit price p and y units of input 2 at unit price q.

The constraint $g(x, y) = u$ tells us that we are only considering the cost of inputs x and y that can be used to produce quantity u of the product.

Thus the number $C(p, q, u)$ gives the minimum cost to the company of producing quantity u if the inputs it needs have unit prices p and q.

(b) The Lagrangian function is

$$\mathcal{L}(x, y, \lambda) = px + qy - \lambda(xy - u).$$

We look for solutions to the system of equations we get from grad $\mathcal{L} = \vec{0}$:

$$\begin{aligned} \frac{\partial \mathcal{L}}{\partial x} &= p - \lambda y = 0 \\ \frac{\partial \mathcal{L}}{\partial y} &= q - \lambda x = 0 \\ \frac{\partial \mathcal{L}}{\partial \lambda} &= -(xy - u) = 0. \end{aligned}$$

We see that $\lambda = p/y = q/x$ so $y = px/q$. Substituting for y in the constraint $xy = u$ leads to $x = \sqrt{qu/p}$, $y = \sqrt{pu/q}$ and $\lambda = \sqrt{pq/u}$. The minimum cost is thus

$$C(p,q,u) = p\sqrt{\frac{qu}{p}} + q\sqrt{\frac{pu}{q}} = 2\sqrt{pqu}.$$

Solutions for Chapter 15 Review

Exercises

1. At a critical point

$$f_x(x,y) = 2xy - 2y = 0$$
$$f_y(x,y) = x^2 + 4y - 2x = 0.$$

From the first equation, $2y(x-1) = 0$, so either $y = 0$ or $x = 1$. If $y = 0$, then $x^2 - 2x = 0$, so $x = 0$ or $x = 2$. Thus $(0,0)$ and $(2,0)$ are critical points. If $x = 1$, then $1^2 + 4y - 2 = 0$, so $y = 1/4$. Thus $(1, 1/4)$ is a critical point. Now

$$D = f_{xx}f_{yy} - (f_{xy})^2 = 2y \cdot 4 - (2x-2)^2 = 8y - 4(x-1)^2,$$

so

$$D(0,0) = -4, \quad D(2,0) = -4, \quad D(1, \tfrac{1}{4}) = 2$$

so $(0,0)$ and $(2,0)$ are saddle points. Since $f_{yy} = 4 > 0$, we see that $(1, 1/4)$ is a local minimum.

5. We find critical points:

$$f_x(x,y) = 12 - 6x = 0$$
$$f_y(x,y) = 6 - 2y = 0$$

so $(2,3)$ is the only critical point. At this point

$$D = f_{xx}f_{yy} - (f_{xy})^2 = (-6)(-2) = 12 > 0,$$

and $f_{xx} < 0$, so $(2,3)$ is a local maximum. Since this is a quadratic, the local maximum is a global maximum.

Alternatively, we complete the square, giving

$$f(x,y) = 10 - 3(x^2 - 4x) - (y^2 - 6y) = 31 - 3(x-2)^2 - (y-3)^2.$$

This expression for f shows that its maximum value (which is 31) occurs where $x = 2, y = 3$.

9. The objective function is $f(x,y) = x^2 + 2y^2$ and the constraint equation is $g(x,y) = 3x + 5y = 200$, so $\text{grad}\, f = (2x)\vec{i} + (4y)\vec{j}$ and $\text{grad}\, g = 3\vec{i} + 5\vec{j}$. Setting $\text{grad}\, f = \lambda\, \text{grad}\, g$ gives

$$2x = 3\lambda,$$
$$4y = 5\lambda.$$

From the first equation, we have $\lambda = 2x/3$, and from the second equation we have $\lambda = 4y/5$. Setting these equal gives

$$x = 1.2y.$$

Substituting this into the constraint equation $3x + 5y = 200$ gives $y = 23.256$. Since $x = 1.2y$, we have $x = 27.907$. A maximum or minimum value of f can occur only at $(27.907, 23.256)$.

We have $f(27.907, 23.256) = 1860.484$. From Figure 15.6, we see that the point $(27.907, 23.256)$ is a minimum value of f subject to the given constraint.

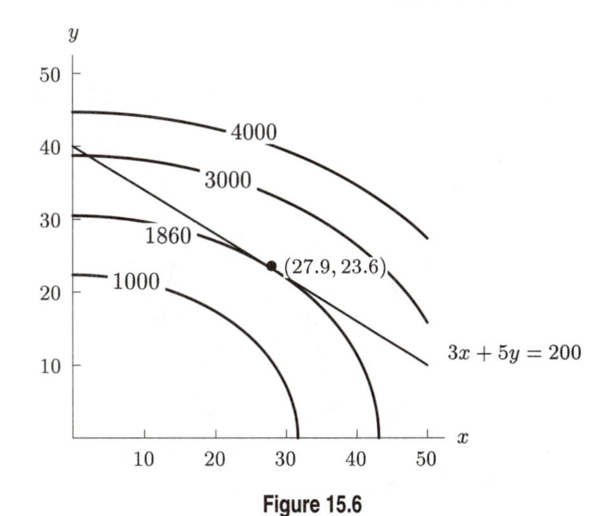

Figure 15.6

13. The region $x^2 \geq y$ is the shaded region in Figure 15.7 which includes the parabola $y = x^2$.

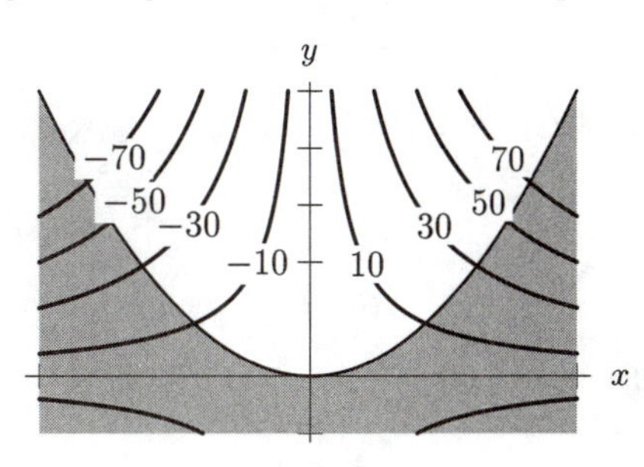

Figure 15.7

We first want to find the local maxima and minima of f in the interior of our region. So we need to find the extrema of

$$f(x, y) = x^2 - y^2, \quad \text{in the region} \quad x^2 > y.$$

For this we compute the critical points:

$$f_x = 2x = 0$$
$$f_y = -2y = 0.$$

As $(0, 0)$ does not belong to the region $x^2 > y$, we have no critical points. Now let's find the local extrema of f on the boundary of our region, hence this time we have to solve a constraint problem. We want to find the extrema of $f(x, y) = x^2 - y^2$ subject to $g(x, y) = x^2 - y = 0$. We use Lagrange multipliers:

$$\operatorname{grad} f = \lambda \operatorname{grad} g \quad \text{and} \quad x^2 = y.$$

This gives

$$2x = 2\lambda x$$
$$2y = \lambda$$
$$x^2 = y.$$

From the first equation we get $x = 0$ or $\lambda = 1$.

If $x = 0$, from the third equation we get $y = 0$, so one solution is $(0, 0)$. If $x \neq 0$, then $\lambda = 1$ and from the second equation we get $y = \frac{1}{2}$. This gives $x^2 = \frac{1}{2}$ so the solutions $(\frac{1}{\sqrt{2}}, \frac{1}{2})$ and $(-\frac{1}{\sqrt{2}}, \frac{1}{2})$.

So $f(0, 0) = 0$ and $f(\frac{1}{\sqrt{2}}, \frac{1}{2}) = f(-\frac{1}{\sqrt{2}}, \frac{1}{2}) = \frac{1}{4}$. From Figure 15.7 showing the level curves of f and the region $x^2 \geq y$, we see that $(0, 0)$ is a local minimum of f on $x^2 = y$, but not a global minimum and that $(\frac{1}{\sqrt{2}}, \frac{1}{2})$ and $(-\frac{1}{\sqrt{2}}, \frac{1}{2})$ are global maxima of f on $x^2 = y$ but *not* global maxima of f on the whole region $x^2 \geq y$.

So there are no global extrema of f in the region $x^2 \geq y$.

Problems

17. Since $f_{xx} < 0$ and $D = f_{xx}f_{yy} - f_{xy}^2 > 0$, the point $(1, 3)$ is a maximum. See Figure 15.8.

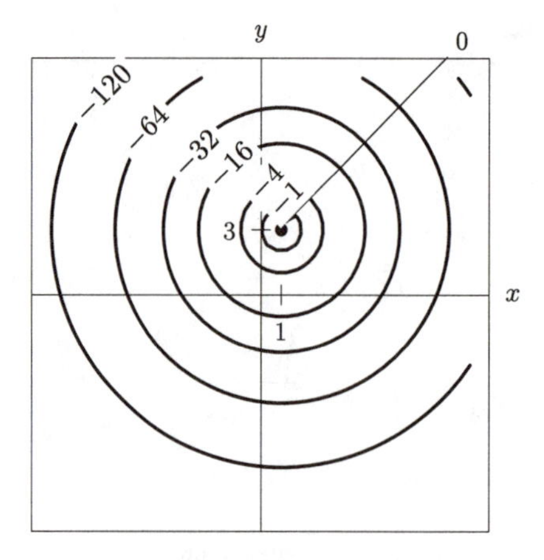

Figure 15.8

21. We want to maximize $f(x, y) = 80x^{0.75}y^{0.25}$ subject to the budget constraint $g(x, y) = 6x + 4y = 8000$. Setting $\operatorname{grad} f = \lambda \operatorname{grad} g$ gives us

$$80(0.75x^{-0.25})y^{0.25} = 6\lambda,$$
$$80x^{0.75}(0.25y^{-0.75}) = 4\lambda.$$

At the maximum, $x, y \neq 0$. From the first equation we have $\lambda = 10y^{0.25}/x^{0.25}$, and from the second equation we have $\lambda = 5x^{0.75}/y^{0.75}$. Setting these equal gives

$$y = 0.5x.$$

Substituting this into the constraint equation $6x+4y = 8000$, we see that $x = 1000$. Since $y = 0.5x$, we obtain $y = 500$. That the point $(1000, 500)$ gives the maximum production is suggested by Figure 15.9, since the values of f decrease as we move along the constraint away from $(1000, 500)$. To produce the maximum quantity, the company should use 1000 units of labor and 500 units of capital.

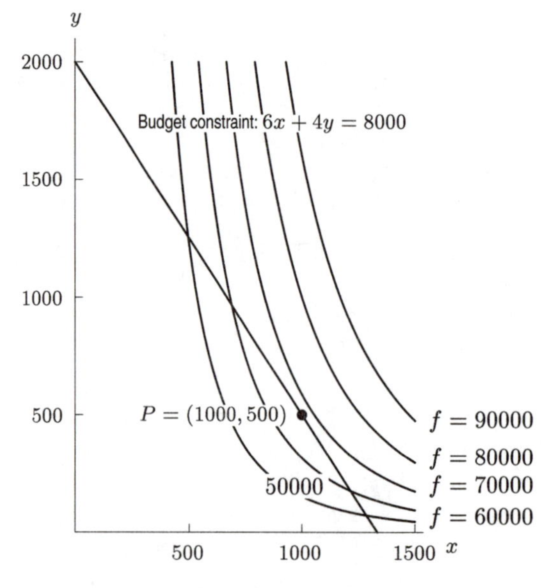

Figure 15.9

25. **(a)** To be producing the maximum quantity Q under the cost constraint given, the firm should be using K and L values given by

$$\frac{\partial Q}{\partial K} = 0.6aK^{-0.4}L^{0.4} = 20\lambda$$
$$\frac{\partial Q}{\partial L} = 0.4aK^{0.6}L^{-0.6} = 10\lambda$$
$$20K + 10L = 150.$$

Hence $\dfrac{0.6aK^{-0.4}L^{0.4}}{0.4aK^{0.6}L^{-0.6}} = 1.5\dfrac{L}{K} = \dfrac{20\lambda}{10\lambda} = 2$, so $L = \dfrac{4}{3}K$. Substituting in $20K + 10L = 150$, we obtain $20K + 10\left(\dfrac{4}{3}\right)K = 150$. Then $K = \dfrac{9}{2}$ and $L = 6$, so capital should be reduced by $\dfrac{1}{2}$ unit, and labor should be increased by 1 unit.

(b) $\dfrac{\text{New production}}{\text{Old production}} = \dfrac{a4.5^{0.6}6^{0.4}}{a5^{0.6}5^{0.4}} \approx 1.01$, so tell the board of directors, "Reducing the quantity of capital by $1/2$ unit and increasing the quantity of labor by 1 unit will increase production by 1% while holding costs to \$150."

29. **(a)** The objective function is the complementary energy, $\dfrac{f_1^2}{2k_1} + \dfrac{f_2^2}{2k_2}$, and the constraint is $f_1 + f_2 = mg$. The Lagrangian function is

$$\mathcal{L}(f_1, f_2, \lambda) = \frac{f_1^2}{2k_1} + \frac{f_2^2}{2k_2} - \lambda(f_1 + f_2 - mg).$$

We look for solutions to the system of equations we get from grad $\mathcal{L} = \vec{0}$:

$$\frac{\partial \mathcal{L}}{\partial f_1} = \frac{f_1}{k_1} - \lambda = 0$$
$$\frac{\partial \mathcal{L}}{\partial f_2} = \frac{f_2}{k_2} - \lambda = 0$$
$$\frac{\partial \mathcal{L}}{\partial \lambda} = -(f_1 + f_2 - mg) = 0.$$

Combining $\dfrac{\partial \mathcal{L}}{\partial f_1} - \dfrac{\partial \mathcal{L}}{\partial f_2} = \dfrac{f_1}{k_1} - \dfrac{f_2}{k_2} = 0$ with $\dfrac{\partial \mathcal{L}}{\partial \lambda} = 0$ gives the two equation system

$$\frac{f_1}{k_1} - \frac{f_2}{k_2} = 0$$
$$f_1 + f_2 = mg.$$

Substituting $f_2 = mg - f_1$ into the first equation leads to

$$f_1 = \frac{k_1}{k_1 + k_2}mg$$
$$f_2 = \frac{k_2}{k_1 + k_2}mg.$$

(b) Hooke's Law states that for a spring

Force of spring = Spring constant · Distance stretched or compressed from equilibrium.

Since $f_1 = k_1 \cdot \lambda$ and $f_2 = k_2 \cdot \lambda$, the Lagrange multiplier λ equals the distance the mass stretches the top spring and compresses the lower spring.

33. The wetted perimeter of the trapezoid is given by the sum of the lengths of the three walls, so

$$p = w + \frac{2d}{\sin\theta}$$

We want to minimize p subject to the constraint that the area is fixed at 50 m^2. A trapezoid of height h and with parallel sides of lengths b_1 and b_2 has

$$A = \text{Area} = h\frac{(b_1 + b_2)}{2}.$$

In this case, d corresponds to h and b_1 corresponds to w. The b_2 term corresponds to the width of the exposed surface of the canal. We find that $b_2 = w + (2d)/(\tan\theta)$. Substituting into our original equation for the area along with the fact that the area is fixed at 50 m^2, we arrive at the formula:

$$\text{Area} = \frac{d}{2}\left(w + w + \frac{2d}{\tan\theta}\right) = d\left(w + \frac{d}{\tan\theta}\right) = 50$$

We now solve the constraint equation for one of the variables; we will choose w to give

$$w = \frac{50}{d} - \frac{d}{\tan\theta}.$$

Substituting into the expression for p gives

$$p = w + \frac{2d}{\sin\theta} = \frac{50}{d} - \frac{d}{\tan\theta} + \frac{2d}{\sin\theta}.$$

We now take partial derivatives:

$$\frac{\partial p}{\partial d} = -\frac{50}{d^2} - \frac{1}{\tan\theta} + \frac{2}{\sin\theta}$$
$$\frac{\partial p}{\partial \theta} = \frac{d}{\tan^2\theta}\cdot\frac{1}{\cos^2\theta} - \frac{2d}{\sin^2\theta}\cdot\cos\theta$$

From $\partial p/\partial\theta = 0$, we get

$$\frac{d\cdot\cos^2\theta}{\sin^2\theta}\cdot\frac{1}{\cos^2\theta} = \frac{2d}{\sin^2\theta}\cdot\cos\theta.$$

Since $\sin\theta \neq 0$ and $\cos\theta \neq 0$, canceling gives

$$1 = 2\cos\theta$$

so

$$\cos\theta = \frac{1}{2}.$$

$$\text{Since}\quad 0 < \theta < \frac{\pi}{2},\quad \text{we get}\quad \theta = \frac{\pi}{3}.$$

Substituting into the equation $\partial p/\partial d = 0$ and solving for d gives:

$$\frac{-50}{d^2} - \frac{1}{\sqrt{3}} + \frac{2}{\sqrt{3}/2} = 0$$

which leads to

$$d = \sqrt{\frac{50}{\sqrt{3}}} \approx 5.37\text{m}.$$

Then

$$w = \frac{50}{d} - \frac{d}{\tan\theta} \approx \frac{50}{5.37} - \frac{5.37}{\sqrt{3}} \approx 6.21 \text{ m}.$$

When $\theta = \pi/3$, $w \approx 6.21$ m and $d \approx 5.37$ m, we have $p \approx 18.61$ m.

Since there is only one critical point, and since p increases without limit as d or θ shrink to zero, the critical point must give the global minimum for p.

CAS Challenge Problems

37. (a) We have grad $f = 3\vec{i} + 2\vec{j}$ and grad $g = (4x - 4y)\vec{i} + (-4x + 10y)\vec{j}$, so the Lagrange multiplier equations are

$$3 = \lambda(4x - 4y)$$
$$2 = \lambda(-4x + 10y)$$
$$2x^2 - 4xy + 5y^2 = 20$$

Solving these with a CAS we get $\lambda = -0.4005, x = -3.9532, y = -2.0806$ and $\lambda = 0.4005, x = 3.9532, y = 2.0806$. We have $f(-3.9532, -2.0806) = -11.0208$, and $f(3,9532, 2.0806) = 21.0208$. The constraint equation is $2x^2 - 4xy + 5y^2 = 20$, or, completing the square, $2(x - y)^2 + 3y^2 = 20$. This has the shape of a skewed ellipse, so the constraint curve is bounded, and therefore the local maximum is a global maximum. Thus the maximum value is 21.0208.

(b) The maximum value on $g = 20.5$ is $\approx 21.0208 + 0.5(0.4005) = 21.2211$. The maximum value on $g = 20.2$ is $\approx 21.0208 + 0.2(0.4005) = 21.1008$.

(c) We use the same commands in the CAS from part (a), with 20 replaced by 20.5 and 20.2, and get the maximum values 21.2198 for $g = 20.5$ and 21.1007 for $g = 20.2$. These agree with the approximations we found in part (b) to 2 decimal places.

CHECK YOUR UNDERSTANDING

1. True. By definition, a critical point is either where the gradient of f is zero or does not exist.

5. True. The graph of this function is a cone that opens upward with its vertex at the origin.

9. False. For example, the linear function $f(x, y) = x + y$ has no local extrema at all.

13. False. For example, the linear function $f(x, y) = x + y$ has neither a global minimum or global maximum on all of 2-space.

17. False. On the given region the function f is always less than one. By picking points closer and closer to the circle $x^2 + y^2 = 1$ we can make f larger and larger (although never larger than one). There is no point in the open disk that gives f its largest value.

21. True. The point (a, b) must lie on the constraint $g(x, y) = c$, so $g(a, b) = c$.

25. False. Since grad f and grad g point in opposite directions, they are parallel. Therefore (a, b) could be a local maximum or local minimum of f constrained to $g = c$. However the information given is not enough to determine that it is a minimum. If the contours of g near (a, b) increase in the opposite direction as the contours of f, then at a point with grad $f(a, b) = \lambda$grad $g(a, b)$ we have $\lambda \leq 0$, but this can be a local maximum or minimum.

For example, $f(x, y) = 4 - x^2 - y^2$ has a local maximum at $(1, 1)$ on the constraint $g(x, y) = x + y = 2$. Yet at this point, grad $f = -2\vec{i} - 2\vec{j}$ and grad $g = \vec{i} + \vec{j}$, so grad f and grad g point in opposite directions.

29. True. Since $f(a, b) = M$, we must satisfy the Lagrange conditions that $f_x(a, b) = \lambda g(a, b)$ and $f_y(a, b) = \lambda g_y(a, b)$, for some λ. Thus $f_x(a, b)/f_y(a, b) = g_x(a, b)/g_y(a, b)$.

33. False. The value of λ at a minimum point gives the proportional change in m for a change in c. If $\lambda > 0$ and the change in c is positive, the change in m will also be positive.

CHAPTER SIXTEEN

Solutions for Section 16.1

Exercises

1. In the subrectangle in the top left in Figure 1, it appears that $f(x, y)$ has a maximum value of about 9. In the subrectangle in the top middle, $f(x, y)$ has a maximum value of 10. Continuing in this way, and multiplying by Δx and Δy, we have

$$\text{Overestimate} = (9 + 10 + 12 + 7 + 8 + 10 + 5 + 7 + 8)(10)(5) = 3800.$$

Similarly, we find

$$\text{Underestimate} = (7 + 7 + 8 + 4 + 5 + 7 + 1 + 3 + 6)(10)(5) = 2400.$$

Thus, we expect that

$$2400 \leq \int_R f(x, y)dA \leq 3800.$$

5. Partition R into subrectangles with the lines $x = 0$, $x = 0.5$, $x = 1$, $x = 1.5$, and $x = 2$ and the lines $y = 0$, $y = 1$, $y = 2$, $y = 3$, and $y = 4$. Then we have 16 subrectangles, each of which we denote $R_{(a,b)}$, where (a, b) is the location of the lower-left corner of the subrectangle.

We want to find a lower bound and an upper bound for the volume above each subrectangle. The lower bound for the volume of $R_{(a,b)}$ is

$$0.5(\text{Min of } f \text{ on } R_{(a,b)})$$

because the area of $R_{(a,b)}$ is $0.5 \cdot 1 = 0.5$. The function $f(x, y) = 2 + xy$ increases with both x and y over the whole region R, as shown in Figure 16.1. Thus,

$$\text{Min of } f \text{ on } R_{(a,b)} = f(a, b) = 2 + ab,$$

because the minimum on each subrectangle is at the corner closest to the origin.

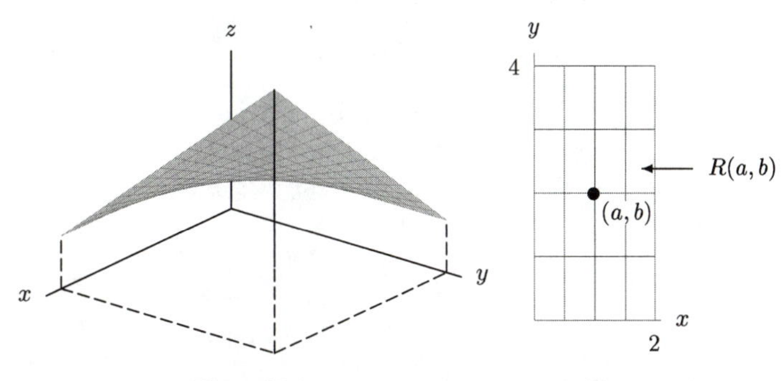

Figure 16.1

Figure 16.2

Similarly,

$$\text{Max of } f \text{ on } R_{(a,b)} = f(a + 0.5, b + 1) = 2 + (a + 0.5)(b + 1).$$

So we have

$$\begin{aligned}\text{Lower sum } &= \sum_{(a,b)} 0.5(2 + ab) = 0.5 \sum_{(a,b)} (2 + ab) \\ &= 16 + 0.5 \sum_{(a,b)} ab\end{aligned}$$

Since $a = 0, 0.5, 1, 1.5$ and $b = 0, 1, 2, 3$, expanding this sum gives

$$\begin{aligned}\text{Lower sum } &= 16 + 0.5\ (\ 0\cdot 0 + 0\cdot 1 + 0\cdot 2 + 0\cdot 3\\ &\quad + 0.5\cdot 0 + 0.5\cdot 1 + 0.5\cdot 2 + 0.5\cdot 3\\ &\quad + 1\cdot 0 + 1\cdot 1 + 1\cdot 2 + 1\cdot 3\\ &\quad + 1.5\cdot 0 + 1.5\cdot 1 + 1.5\cdot 2 + 1.5\cdot 3)\\ &= 25.\end{aligned}$$

Similarly, we can compute the upper sum:

$$\begin{aligned}\text{Upper sum } &= \sum_{(a,b)} 0.5(2 + (a+0.5)(b+1)) = 0.5 \sum_{(a,b)} (2 + (a+0.5)(b+1))\\ &= 16 + 0.5 \sum_{(a,b)} (a+0.5)(b+1)\\ &= 41.\end{aligned}$$

Problems

9. Let's break up the room into 25 sections, each of which is 1 meter by 1 meter and has area $\Delta A = 1$.

We shall begin our sum as an upper estimate starting with the lower left corner of the room and continue across the bottom and moving upwards using the highest temperature, T_i, in each case. So the upper Riemann sum becomes

$$\begin{aligned}\textstyle\sum_{i=1}^{25} T_i \Delta A &= T_1\Delta A + T_2\Delta A + T_3\Delta A + \cdots + T_{25}\Delta A\\ &= \Delta A(T_1 + T_2 + T_3 + \cdots + T_{25})\\ &= (1)\ (31 + 29 + 28 + 27 + 27 +\\ &\quad 29 + 28 + 27 + 27 + 26 +\\ &\quad 27 + 27 + 26 + 26 + 26 +\\ &\quad 26 + 26 + 25 + 25 + 25 +\\ &\quad 25 + 24 + 24 + 24 + 24)\\ &= (1)(659) = 659.\end{aligned}$$

In the same way, the lower Riemann sum is formed by taking the lowest temperature, t_i, in each case:

$$\begin{aligned}\textstyle\sum_{i=1}^{25} t_i \Delta A &= t_1\Delta A + t_2\Delta A + t_3\Delta A + \cdots + t_{25}\Delta A\\ &= \Delta A(t_1 + t_2 + t_3 + \cdots + t_{25})\\ &= (1)\ (27 + 27 + 26 + 26 + 25 +\\ &\quad 26 + 26 + 25 + 25 + 25 +\\ &\quad 25 + 24 + 24 + 24 + 24 +\\ &\quad 24 + 23 + 23 + 23 + 23 +\\ &\quad 23 + 21 + 20 + 21 + 22)\\ &= (1)(602) = 602.\end{aligned}$$

So, averaging the upper and lower sums we get: 630.5.

To compute the average temperature, we divide by the area of the room, giving

$$\text{Average temperature} = \frac{630.5}{(5)(5)} \approx 25.2^\circ\text{C}.$$

Alternatively we can use the temperature at the central point of each section ΔA. Then the sum becomes

$$\begin{aligned}\sum_{i=1}^{25} T_i'\Delta A = \Delta A\sum_{i=1}^{25} T_i' &= (1)(29 + 28 + 27 + 26.5 + 26 + \\ &\quad 27 + 27 + 26 + 26 + 25.5 + \\ &\quad 26 + 25.5 + 25 + 25 + 25 + \\ &\quad 25 + 24 + 24 + 24 + 24 + \\ &\quad 24 + 23 + 22 + 22.5 + 23) \\ &= (1)(630) = 630.\end{aligned}$$

Then we get

$$\text{Average temperature} = \frac{\sum_{i=1}^{25} T_i'\Delta A}{\text{Area}} = \frac{630}{(5)(5)} \approx 25.2^\circ\text{C}.$$

13. We use four subrectangles to find an overestimate and underestimate of the integral:

$$\text{Overestimate} = (15 + 9 + 9 + 5)(4)(3) = 456,$$

$$\text{Underestimate} = (5 + 2 + 3 + 1)(4)(3) = 132.$$

A better estimate of the integral is the average of the two:

$$\int_R f(x, y)dA \approx \frac{456 + 132}{2} = 294.$$

The units of the integral are milligrams, and the integral represents the total number of mg of mosquito larvae in this 8 meter by 6 meter section of swamp.

17. The function being integrated is $f(x, y) = 5x$, which is an odd function in x. Since B is symmetric with respect to x, the contributions to the integral cancel out, as $f(x, y) = -f(-x, y)$. Thus, the integral is zero.

21. The region R is symmetric with respect to y and the integrand is an odd function in y, so the integral over R is zero.

25. Since D is a disk of radius 1, in the region D, we have $|y| < 1$. Thus, $-\pi/2 < y < \pi/2$. Thus, $\cos y$ is always positive in the region D and thus its integral is positive.

29. The function $f(x, y)$ is odd with respect to x, and thus the integral is zero in region B, which is symmetric with respect to x.

33. Take a Riemannian sum approximation to

$$\int_R f dA \approx \sum_{i,j} f(x_i, y_i)\Delta A.$$

Then, using the fact that $|a + b| \le |a| + |b|$ repeatedly, we have:

$$\left|\int_R f dA\right| \approx |\sum_{i,j} f(x_i, y_i)\Delta A| \le \sum |f(x_i, y_i)\Delta A|.$$

Now $|f(x_i, y_j)\Delta A| = |f(x_i, y_j|\Delta A$ since ΔA is non-negative, so

$$\left|\int_R f dA\right| \le \sum_{i,j} |f(x_i, y_i)\Delta A| = \sum_{i,j} |f(x_i, y_j)|\Delta A.$$

But the last expression on the right is a Riemann sum approximation to the integral $\int_R |f|dA$, so we have

$$\left|\int_R f dA\right| \approx \left|\sum_{i,j} f(x_i, y_j)\Delta A\right| \le \sum_{i,j} |f(x_{i,j})|\Delta A \approx \int_R |f|dA.$$

Thus,

$$\left|\int_R f dA\right| \le \int_R |f|dA.$$

Solutions for Section 16.2

Exercises

1. We evaluate the inside integral first:

$$\int_0^3 (x^2 + y^2)\, dy = \left(x^2 y + \frac{y^3}{3}\right)\Bigg|_{y=0}^{y=3} = 3x^2 + 9.$$

Therefore, we have

$$\int_0^2 \int_0^3 (x^2 + y^2)\, dydx = \int_0^2 (3x^2 + 9)\, dx = (x^3 + 9x)\Big|_0^2 = 26.$$

5. $\displaystyle\int_1^4 \int_1^2 f\, dy\, dx \quad \text{or} \quad \int_1^2 \int_1^4 f\, dx\, dy$

9. The line connecting $(1, 0)$ and $(4, 1)$ is

$$y = \frac{1}{3}(x - 1)$$

So the integral is

$$\int_1^4 \int_{(x-1)/3}^2 f\, dy\, dx$$

13.

$$\begin{aligned}
\int_1^5 \int_x^{2x} \sin x\, dy\, dx &= \int_1^5 \sin x \cdot y\Big|_x^{2x} dx \\
&= \int_1^5 \sin x \cdot x\, dx \\
&= (\sin x - x\cos x)\Big|_1^5 \\
&= (\sin 5 - 5\cos 5) - (\sin 1 - \cos 1) \approx -2.68.
\end{aligned}$$

See Figure 16.3.

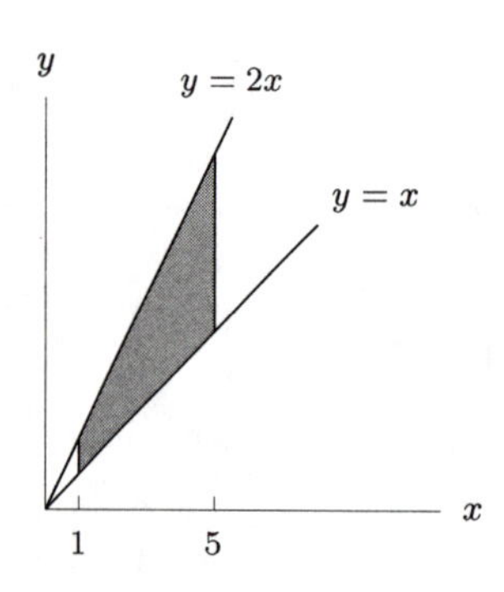

Figure 16.3

17. The region of integration, R, is shown in Figure 16.4.

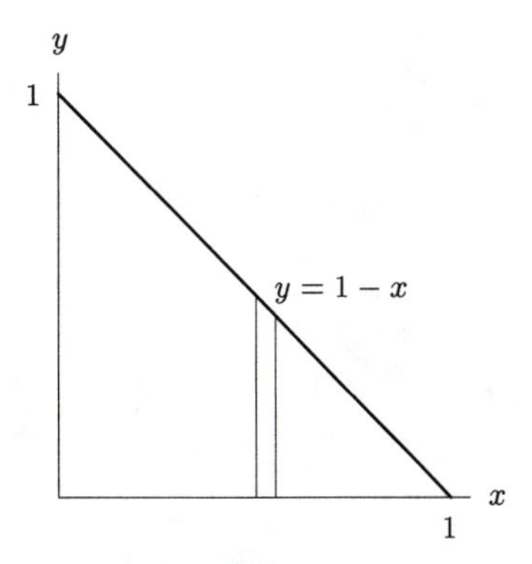

Figure 16.4

Integrating first over y, as shown in the diagram, we obtain

$$\int_R xy\,dA = \int_0^1 \left(\int_0^{1-x} xy\,dy\right) dx = \int_0^1 \left.\frac{xy^2}{2}\right|_0^{1-x} dx = \int_0^1 \frac{1}{2}x(1-x)^2\,dx$$

Now integrating with respect to x gives

$$\int_R xy\,dA = \left.\left(\frac{1}{4}x^2 - \frac{1}{3}x^3 + \frac{1}{8}x^4\right)\right|_0^1 = \frac{1}{24}.$$

21. It would be easier to integrate first in the x direction from $x = y - 1$ to $x = -y + 1$, because integrating first in the y direction would involve two separate integrals.

$$\begin{aligned}
\int_R (2x+3y)^2\,dA &= \int_0^1 \int_{y-1}^{-y+1} (2x+3y)^2\,dx\,dy \\
&= \int_0^1 \int_{y-1}^{-y+1} (4x^2 + 12xy + 9y^2)\,dx\,dy \\
&= \int_0^1 \left[\frac{4}{3}x^3 + 6x^2y + 9xy^2\right]_{y-1}^{-y+1} dy \\
&= \int_0^1 [\frac{8}{3}(-y+1)^3 + 9y^2(-2y+2)]\,dy \\
&= \left[-\frac{2}{3}(-y+1)^4 - \frac{9}{2}y^4 + 6y^3\right]_0^1 \\
&= -\frac{2}{3}(-1) - \frac{9}{2} + 6 = \frac{13}{6}
\end{aligned}$$

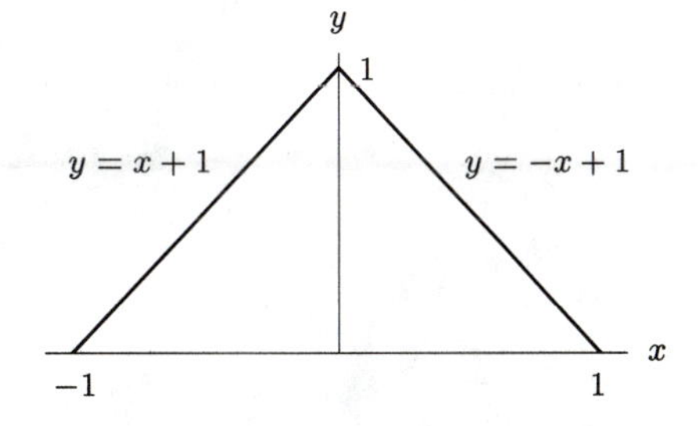

Figure 16.5

25. **(a)**

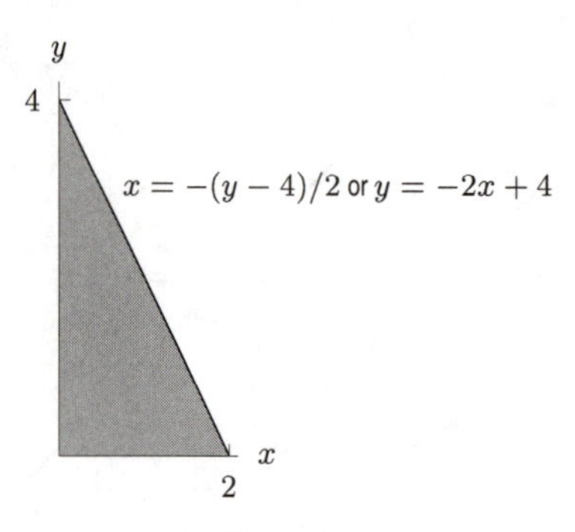

Figure 16.6

(b) $\int_0^2 \int_0^{-2x+4} g(x, y)\, dy\, dx$.

29. As given, the region of integration is as shown in Figure 16.7.

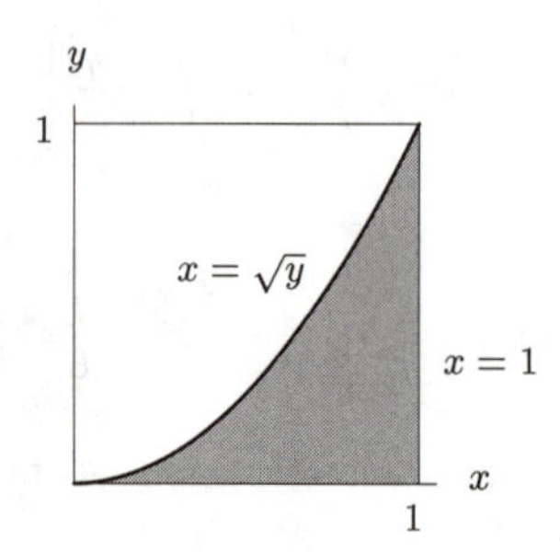

Figure 16.7

Reversing the limits gives

$$\begin{aligned}\int_0^1 \int_0^{x^2} \sqrt{2 + x^3}\, dy dx &= \int_0^1 (y\sqrt{2 + x^3}\Big|_0^{x^2})\, dx \\ &= \int_0^1 x^2 \sqrt{2 + x^3}\, dx \\ &= \frac{2}{9}(2 + x^3)^{\frac{3}{2}}\Big|_0^1 = \frac{2}{9}(3\sqrt{3} - 2\sqrt{2}).\end{aligned}$$

Problems

33. The intersection of the graph of $f(x, y) = 25 - x^2 - y^2$ and xy-plane is a circle $x^2 + y^2 = 25$. The given solid is shown in Figure 16.8.

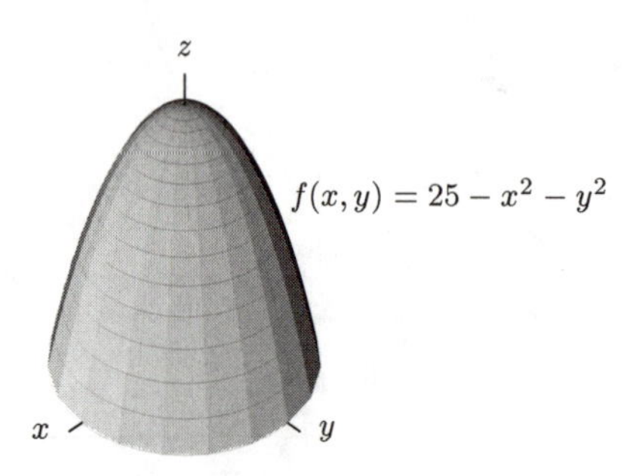

Figure 16.8

Thus the volume of the solid is

$$V = \int_R f(x,y)\,dA$$
$$= \int_{-5}^{5}\int_{-\sqrt{25-y^2}}^{\sqrt{25-y^2}} (25 - x^2 - y^2)\,dx\,dy.$$

37. The region of integration is shown in Figure 16.9. Thus

$$\text{Volume} = \int_0^1\int_0^x (x^2+y^2)\,dy\,dx = \int_0^1 \left(x^2y + \frac{y^3}{3}\right)\Bigg|_{y=0}^{y=x} dx = \int_0^1 \frac{4}{3}x^3\,dx = \frac{x^4}{3}\Bigg|_0^1 = \frac{1}{3}.$$

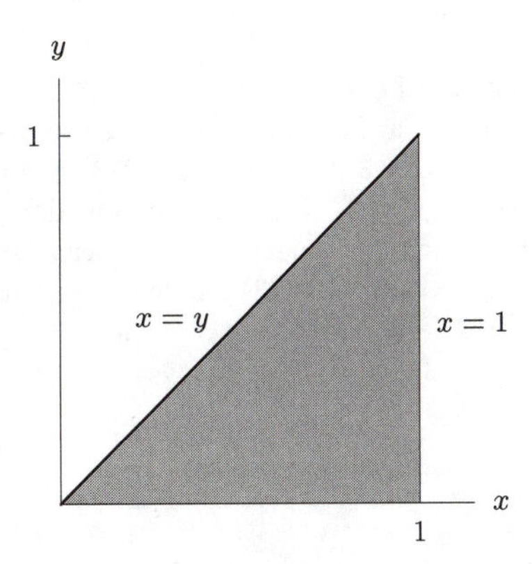

Figure 16.9

41. We want to calculate the volume of the tetrahedron shown in Figure 16.10.

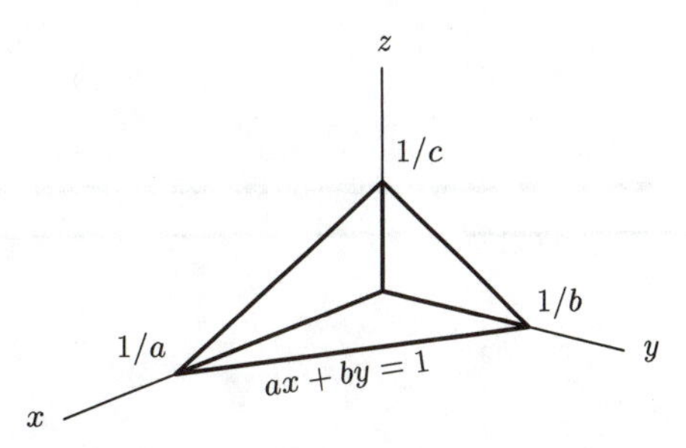

Figure 16.10

We first find the region in the xy-plane where the graph of $ax + by + cz = 1$ is above the xy-plane. When $z = 0$ we have $ax + by = 1$. So the region over which we want to integrate is bounded by $x = 0, y = 0$ and $ax + by = 1$. Integrating with respect to y first, we have

$$\text{Volume} = \int_0^{1/a}\int_0^{(1-ax)/b} z\,dy\,dx = \int_0^{1/a}\int_0^{(1-ax)/b} \frac{1-by-ax}{c}\,dy\,dx$$
$$= \int_0^{1/a}\left(\frac{y}{c} - \frac{by^2}{2c} - \frac{axy}{c}\right)\Bigg|_{y=0}^{y=(1-ax)/b} dx$$
$$= \int_0^{1/a} \frac{1}{2bc}(1 - 2ax + a^2x^2)\,dx$$
$$= \frac{1}{6abc}.$$

45. **(a)** We have

$$\begin{aligned}
\text{Average value of } f &= \frac{1}{\text{Area of Rectangle}} \int_{\text{Rectangle}} f\, dA \\
&= \frac{1}{6}\int_{x=0}^{2}\int_{y=0}^{3} (ax+by)\, dydx = \frac{1}{6}\int_0^2 \left(axy + b\frac{y^2}{2}\right)\Bigg|_{y=0}^{y=3} dx \\
&= \frac{1}{6}\int_0^2 \left(3ax + \frac{9}{2}b\right) dx = \frac{1}{6}\left(\frac{3}{2}ax^2 + \frac{9}{2}bx\right)\Bigg|_0^2 \\
&= \frac{1}{6}(6a+9b) \\
&= a + \frac{3}{2}b.
\end{aligned}$$

The average value will be 20 if and only if $a + (3/2)b = 20$.

This equation can also be expressed as $2a + 3b = 40$, which shows that $f(x, y) = ax + by$ has average value of 20 on the rectangle $0 \le x \le 2, 0 \le y \le 3$ if and only if $f(2, 3) = 40$.

(b) Since $2a + 3b = 40$, we must have $b = (40/3) - (2/3)a$. Any function $f(x, y) = ax + ((40/3) - (2/3)a)y$ where a is any real number is a correct solution. For example, $a = 1$ leads to the function $f(x, y) = x + (38/3)y$, and $a = -3$ leads to the function $f(x, y) = -3x + (46/3)y$, both of which have average value 20 on the given rectangle. See Figure 16.11 and 16.12.

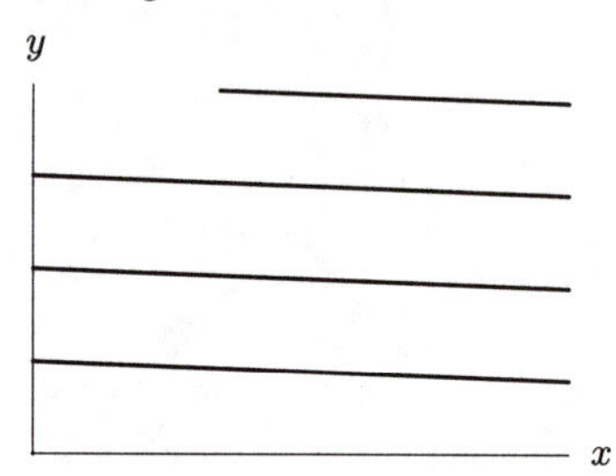

Figure 16.11: $f(x, y) = x + \frac{38}{3}y$

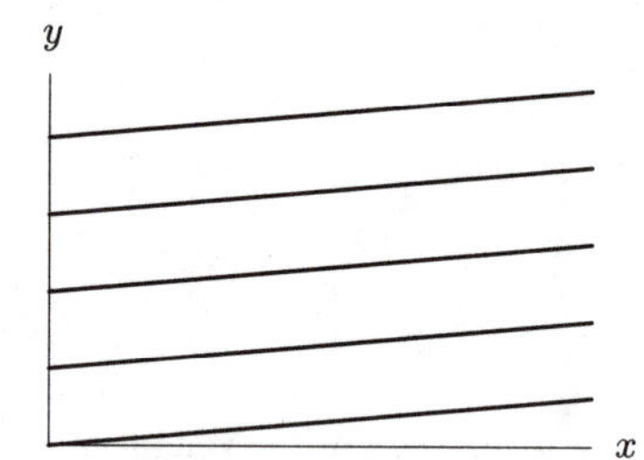

Figure 16.12: $f(x, y) = -3x + \frac{46}{3}y$

Solutions for Section 16.3

Exercises

1.

$$\begin{aligned}
\int_W f\, dV &= \int_0^2\int_{-1}^1\int_2^3 (x^2 + 5y^2 - z)\, dz\, dy\, dx \\
&= \int_0^2\int_{-1}^1 (x^2z + 5y^2z - \frac{1}{2}z^2)\Big|_2^3 dy\, dx \\
&= \int_0^2\int_{-1}^1 (x^2 + 5y^2 - \frac{5}{2})\, dy\, dx \\
&= \int_0^2 (x^2y + \frac{5}{3}y^3 - \frac{5}{2}y)\Big|_{-1}^1 dx \\
&= \int_0^2 (2x^2 + \frac{10}{3} - 5)\, dx \\
&= (\frac{2}{3}x^3 - \frac{5}{3}x)\Big|_0^2 \\
&= \frac{16}{3} - \frac{10}{3} = 2
\end{aligned}$$

5. The region is the half cylinder in Figure 16.13.

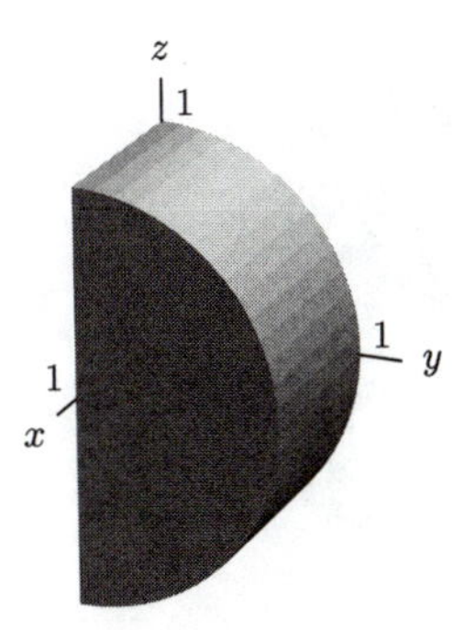

Figure 16.13

9. The region is the half cylinder in Figure 16.14.

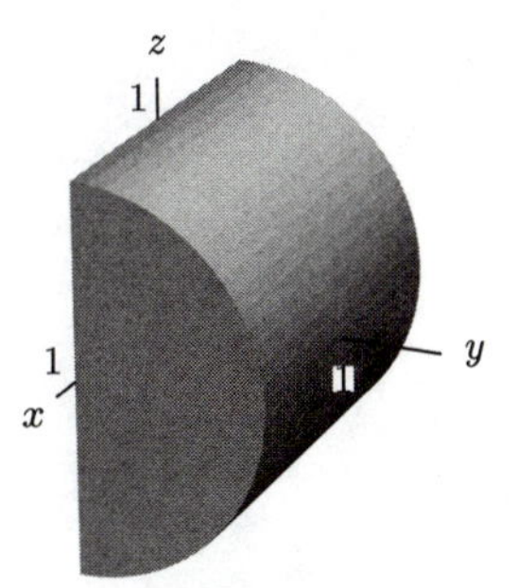

Figure 16.14

13. The region is the quarter sphere in Figure 16.15.

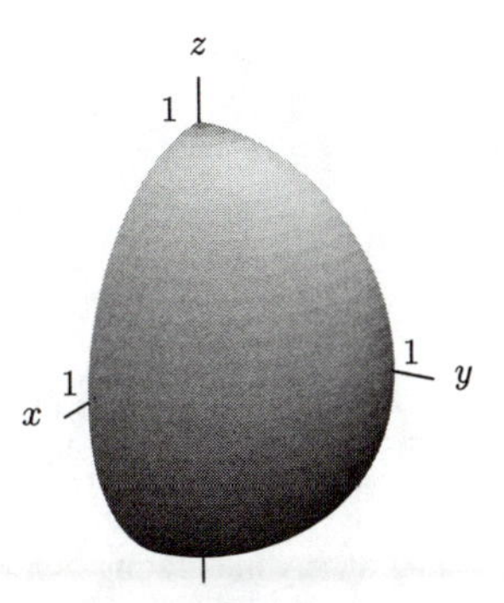

Figure 16.15

Problems

17. The region of integration is shown in Figure 16.16, and the mass of the given solid is given by

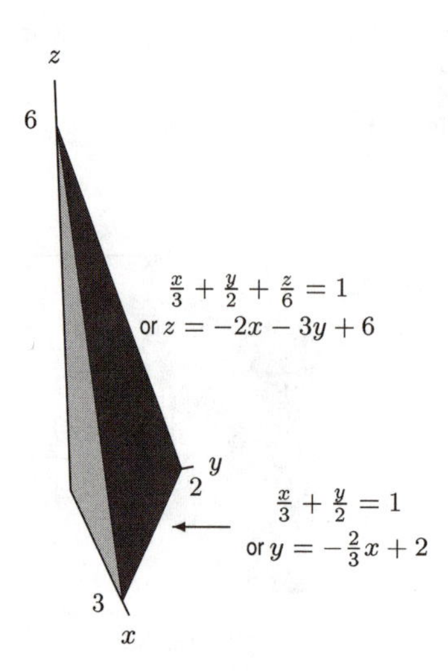

Figure 16.16

$$
\begin{aligned}
\text{mass} &= \int_R \delta\, dV \\
&= \int_0^3 \int_0^{-\frac{2}{3}x+2} \int_0^{-2x-3y+6} (x+y)\, dzdydx \\
&= \int_0^3 \int_0^{-\frac{2}{3}x+2} (x+y)z \Big|_0^{-2x-3y+6} dydx \\
&= \int_0^3 \int_0^{-\frac{2}{3}x+2} (x+y)(-2x-3y+6)\, dydx \\
&= \int_0^3 \int_0^{-\frac{2}{3}x+2} (-2x^2-3y^2-5xy+6x+6y)\, dydx \\
&= \int_0^3 \left(-2x^2y - y^3 - \frac{5}{2}xy^2 + 6xy + 3y^2\right)\Big|_0^{-\frac{2}{3}x+2} dx \\
&= \int_0^3 \left(\frac{14}{27}x^3 - \frac{8}{3}x^2 + 2x + 4\right) dx \\
&= \left(\frac{7}{54}x^4 - \frac{8}{9}x^3 + x^2 + 4x\right)\Big|_0^3 \\
&= \frac{7}{54}\cdot 3^4 - \frac{8}{9}\cdot 3^3 + 3^2 + 12 = \frac{21}{2} - 3 = \frac{15}{2}.
\end{aligned}
$$

21. Zero. The value of x is positive above the first and fourth quadrants in the xy-plane, and negative (and of equal absolute value) above the second and third quadrants. The integral of x over the entire solid cone is zero because the integrals over the two halves of the cone cancel.

25. Zero. Write the triple integral as an iterated integral, say integrating first with respect to x. For fixed y and z, the x-integral is over an interval symmetric about 0. The integral of x over such an interval is zero. If any of the inner integrals in an iterated integral is zero, then the triple integral is zero.

29. Positive. If (x, y, z) is any point inside the solid W then $\sqrt{x^2+y^2} < z$. Thus $z - \sqrt{x^2+y^2} > 0$, and so its integral over the solid W is positive.

33. Zero. You can see this in several ways. One way is to observe that xy is positive on part of the cone above the first quadrant (where x and y are of the same sign) and negative (of equal absolute value) on the part of the cone above the fourth quadrant (where x and y have opposite signs). These add up to zero in the integral of xy over all of W.

Another way to see that the integral is zero is to write the triple integral as an iterated integral, say integrating first with respect to y. For fixed x and z, the y-integral is over an interval symmetric about 0. The integral of y over such an interval is zero. If any of the inner integrals in an iterated integral is zero, then the triple integral is zero.

37. Orient the region as shown in Figure 16.17 and use Cartesian coordinates with origin at the center of the sphere. The equation of the sphere is $x^2 + y^2 + z^2 = 25$, and we want the volume between the planes $z = 3$ and $z = 5$. The plane $z = 3$ cuts the sphere in the circle $x^2 + y^2 + 3^2 = 25$, or $x^2 + y^2 = 16$.

$$\text{Volume} = \int_{-4}^{4} \int_{-\sqrt{16-x^2}}^{\sqrt{16-x^2}} \int_{3}^{\sqrt{25-x^2-y^2}} dzdydx.$$

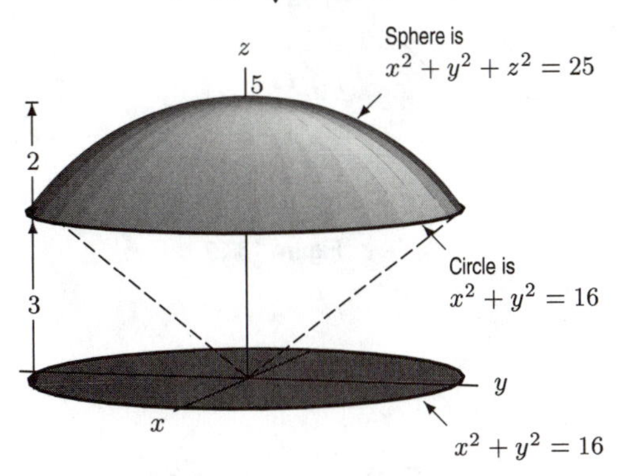

Figure 16.17

41. The volume V of the solid is $1 \cdot 2 \cdot 3 = 6$. We need to compute

$$\begin{aligned}
\frac{m}{6} \int_W x^2 + y^2 \, dV &= \frac{m}{6} \int_0^1 \int_0^2 \int_0^3 x^2 + y^2 \, dz \, dy \, dx \\
&= \frac{m}{6} \int_0^1 \int_0^2 3(x^2 + y^2) \, dy \, dx \\
&= \frac{m}{2} \int_0^1 (x^2 y + y^3/3)\Big|_0^2 \, dx \\
&= \frac{m}{2} \int_0^1 (2x^2 + 8/3) \, dx = 5m/3
\end{aligned}$$

Solutions for Section 16.4

Exercises

1. $\displaystyle\int_{\pi/4}^{3\pi/4} \int_0^2 f \, r dr \, d\theta$

5.

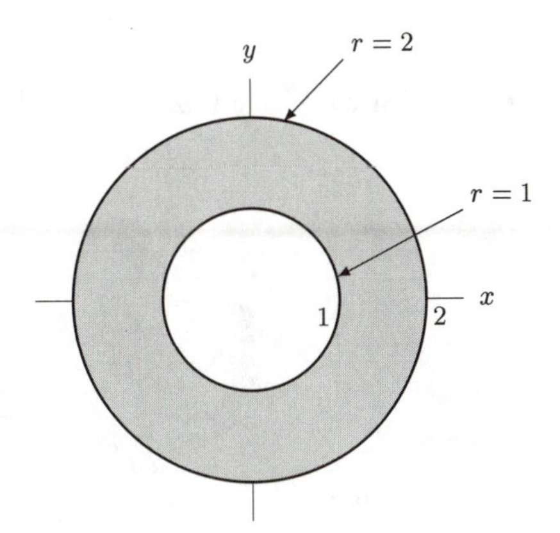

Figure 16.18

9.

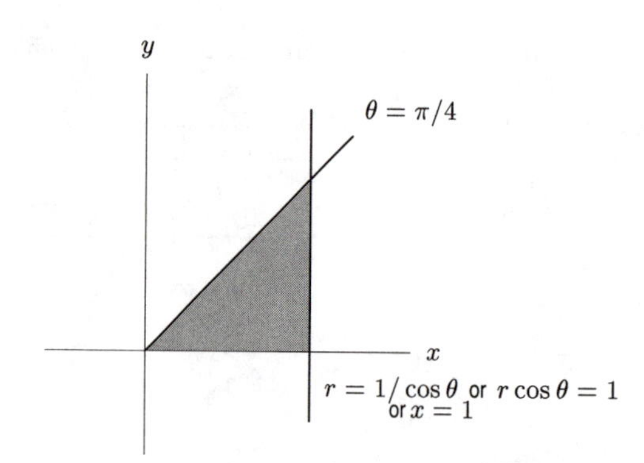

Figure 16.19

13. The region is pictured in Figure 16.20.

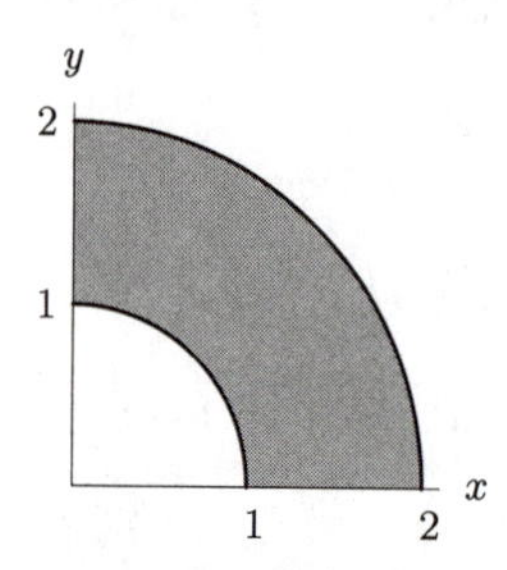

Figure 16.20

Using polar coordinates, we get

$$\int_R (x^2 - y^2) dA = \int_0^{\pi/2} \int_1^2 r^2(\cos^2\theta - \sin^2\theta) r dr\, d\theta = \int_0^{\pi/2} (\cos^2\theta - \sin^2\theta) \cdot \frac{1}{4} r^4 \Big|_1^2 d\theta$$

$$= \frac{15}{4} \int_0^{\pi/2} (\cos^2\theta - \sin^2\theta)\, d\theta$$

$$= \frac{15}{4} \int_0^{\pi/2} \cos 2\theta\, d\theta$$

$$= \frac{15}{4} \cdot \frac{1}{2} \sin 2\theta \Big|_0^{\pi/2} = 0.$$

17. From the given limits, the region of integration is in Figure 16.21.

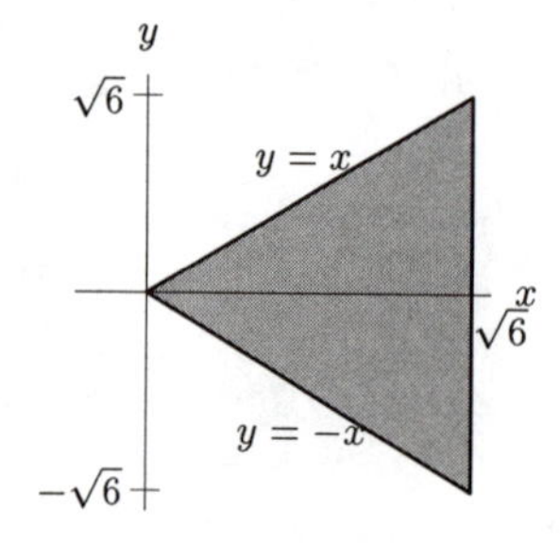

Figure 16.21

In polar coordinates, $-\pi/4 \le \theta \le \pi/4$. Also, $\sqrt{6} = x = r\cos\theta$. Hence, $0 \le r \le \sqrt{6}/\cos\theta$. The integral becomes

$$\begin{aligned}
\int_0^{\sqrt{6}}\int_{-x}^{x} dy\,dx &= \int_{-\pi/4}^{\pi/4}\int_0^{\sqrt{6}/\cos\theta} r\,dr\,d\theta \\
&= \int_{-\pi/4}^{\pi/4}\left(\frac{r^2}{2}\bigg|_0^{\sqrt{6}/\cos\theta}\right) d\theta = \int_{-\pi/4}^{\pi/4}\frac{6}{2\cos^2\theta}\,d\theta \\
&= 3\tan\theta\bigg|_{-\pi/4}^{\pi/4} = 3\cdot(1-(-1)) = 6.
\end{aligned}$$

Notice that we can check this answer because the integral gives the area of the shaded triangular region which is $\frac{1}{2}\cdot\sqrt{6}\cdot(2\sqrt{6}) = 6$.

Problems

21. First, let's find where the two surfaces intersect.

$$\begin{aligned}
\sqrt{8-x^2-y^2} &= \sqrt{x^2+y^2} \\
8-x^2-y^2 &= x^2+y^2 \\
x^2+y^2 &= 4
\end{aligned}$$

So $z = 2$ at the intersection. See Figure 16.22.

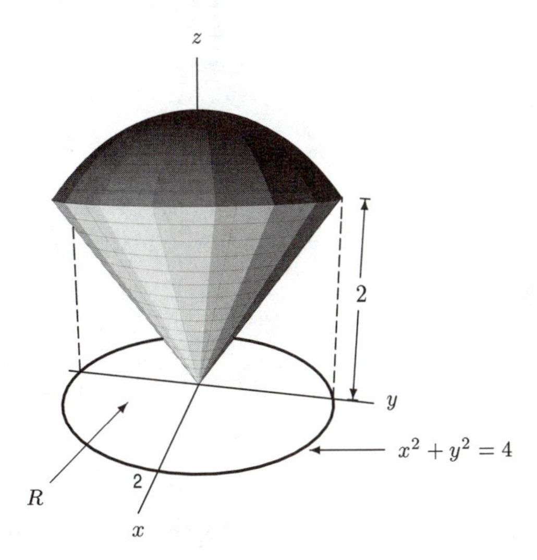

Figure 16.22

The volume of the ice cream cone has two parts. The first part (which is the volume of the cone) is the volume of the solid bounded by the plane $z = 2$ and the cone $z = \sqrt{x^2+y^2}$. Hence, this volume is given by $\int_R (2-\sqrt{x^2+y^2})\,dA$, where R is the disk of radius 2 centered at the origin, in the xy-plane. Using polar coordinates, we have:

$$\begin{aligned}
\int_R\left(2-\sqrt{x^2+y^2}\right)dA &= \int_0^{2\pi}\int_0^2 (2-r)\cdot r\,dr\,d\theta \\
&= \int_0^{2\pi}\left[\left(r^2-\frac{r^3}{3}\right)\bigg|_0^2\right]d\theta \\
&= \frac{4}{3}\int_0^{2\pi} d\theta \\
&= 8\pi/3
\end{aligned}$$

The second part is the volume of the region above the plane $z = 2$ but inside the sphere $x^2 + y^2 + z^2 = 8$, which is given by $\int_R (\sqrt{8 - x^2 - y^2} - 2)\, dA$ where R is the same disk as before. Now

$$\begin{aligned}
\int_R (\sqrt{8 - x^2 - y^2} - 2)\, dA &= \int_0^{2\pi} \int_0^2 (\sqrt{8 - r^2} - 2) r dr\, d\theta \\
&= \int_0^{2\pi} \int_0^2 r\sqrt{8 - r^2}\, dr\, d\theta - \int_0^{2\pi} \int_0^2 2r\, dr\, d\theta \\
&= \int_0^{2\pi} \left(-\frac{1}{3}(8 - r^2)^{3/2} \Big|_0^2 \right) d\theta - \int_0^{2\pi} r^2 \Big|_0^2 d\theta \\
&= -\frac{1}{3} \int_0^{2\pi} (4^{3/2} - 8^{3/2})\, d\theta - \int_0^{2\pi} 4\, d\theta \\
&= -\frac{1}{3} \cdot 2\pi (8 - 16\sqrt{2}) - 8\pi \\
&= \frac{2\pi}{3}(16\sqrt{2} - 8) - 8\pi \\
&= \frac{8\pi(4\sqrt{2} - 5)}{3}
\end{aligned}$$

Thus, the total volume is the sum of the two volumes, which is $32\pi(\sqrt{2} - 1)/3$.

25. **(a)** We must first decide where to put the origin. We locate the origin at the center of one disk and locate the center of the second disk at the point $(1, 0)$. See Figure 16.23. (Other choices of origin are possible.)

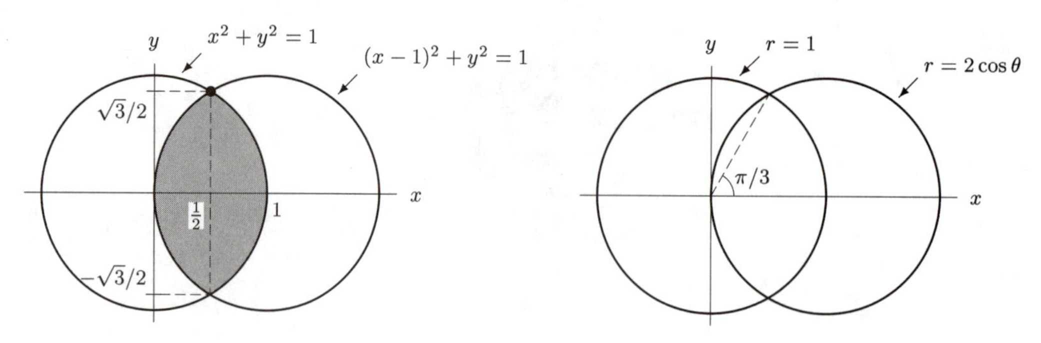

Figure 16.23

Figure 16.24

By symmetry, the points of intersection of the circles are half-way between the centers, at $x = 1/2$. The y-values at these points are given by

$$y = \pm\sqrt{1 - x^2} = \pm\sqrt{1 - \left(\frac{1}{2}\right)^2} = \pm\frac{\sqrt{3}}{2}.$$

We integrate in the x-direction first, so that it is not necessary to set up two integrals. The right-side of the circle $x^2 + y^2 = 1$ is given by

$$x = \sqrt{1 - y^2}.$$

The left side of the circle $(x - 1)^2 + y^2 = 1$ is given by

$$x = 1 - \sqrt{1 - y^2}.$$

Thus the area of overlap is given by

$$\text{Area} = \int_{-\sqrt{3}/2}^{\sqrt{3}/2} \int_{1-\sqrt{1-y^2}}^{\sqrt{1-y^2}} dx dy.$$

(b) In polar coordinates, the circle centered at the origin has equation $r = 1$. See Figure 16.24. The other circle, $(x-1)^2 + y^2 = 1$, can be written as

$$x^2 - 2x + 1 + y^2 = 1$$
$$x^2 + y^2 = 2x,$$

so its equation in polar coordinates is

$$r^2 = 2r\cos\theta,$$

and, since $r \neq 0$,

$$r = 2\cos\theta.$$

At the top point of intersection of the two circles, $x = 1/2$, $y = \sqrt{3}/2$, so $\tan\theta = \sqrt{3}$, giving $\theta = \pi/3$.

Figure 16.24 shows that if we integrate with respect to r first, we have to write the integral as the sum of two integrals. Thus, we integrate with respect to θ first. To do this, we rewrite

$$r = 2\cos\theta \quad \text{as} \quad \theta = \arccos\left(\frac{r}{2}\right).$$

This gives the top half of the circle; the bottom half is given by

$$\theta = -\arccos\left(\frac{r}{2}\right).$$

Thus the area is given by

$$\text{Area} = \int_0^1 \int_{-\arccos(r/2)}^{\arccos(r/2)} r\, d\theta dr.$$

Solutions for Section 16.5

Exercises

1.

$$\int_W f\, dV = \int_{-1}^{1} \int_{\pi/4}^{3\pi/4} \int_0^4 (r^2 + z^2)\, r dr\, d\theta\, dz$$
$$= \int_{-1}^{1} \int_{\pi/4}^{3\pi/4} (64 + 8z^2)\, d\theta\, dz$$
$$= \int_{-1}^{1} \frac{\pi}{2}(64 + 8z^2)\, dz$$
$$= 64\pi + \frac{8}{3}\pi = \frac{200}{3}\pi$$

5. Using Cartesian coordinates, we get:

$$\int_0^3 \int_0^1 \int_0^5 f\, dz\, dy\, dx$$

9. Using spherical coordinates, we get:

$$\int_0^{2\pi} \int_0^{\pi/6} \int_0^3 f \cdot \rho^2 \sin\phi\, d\rho\, d\phi\, d\theta$$

13. **(a)** The region of integration is the region between the cone $z = r$, the xy-plane and the cylinder $r = 3$. In spherical coordinates, $r = 3$ becomes $\rho \sin\phi = 3$, so $\rho = 3/\sin\phi$. The cone is $\phi = \pi/4$ and the xy-plane is $\phi = \pi/2$. See Figure 16.25. Thus, the integral becomes

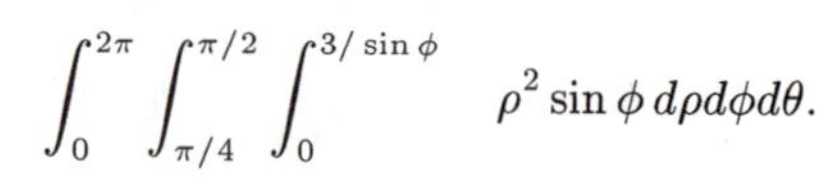

$$\int_0^{2\pi}\int_{\pi/4}^{\pi/2}\int_0^{3/\sin\phi} \rho^2 \sin\phi \, d\rho d\phi d\theta.$$

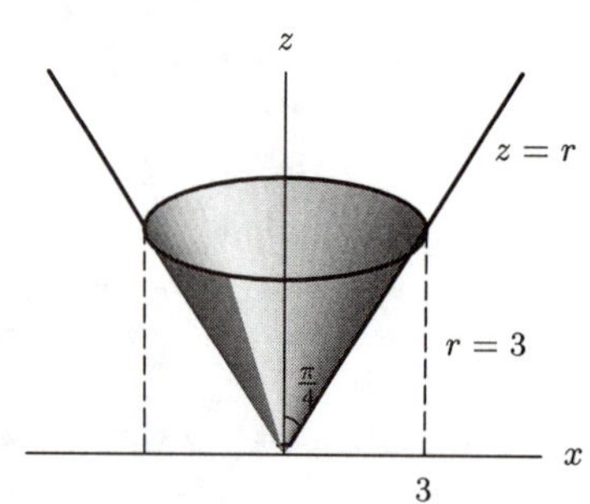

Figure 16.25: Region of integration is between the cone and the xy-plane

(b) The original integral is easier to evaluate, so

$$\int_0^{2\pi}\int_0^3\int_0^r r\,dz dr d\theta = \int_0^{2\pi}\int_0^3 zr\Big|_{z=0}^{z=r} dr d\theta = \int_0^{2\pi}\int_0^3 r^2\,dr d\theta = 2\pi\cdot\frac{r^3}{3}\Big|_0^3 = 18\pi.$$

Problems

17. **(a)** The angle ϕ takes on values in the range $0 \le \phi \le \pi$. Thus, $\sin\phi$ is nonnegative everywhere in W_1, and so its integral is positive.

(b) The function ϕ is symmetric across the xy plane, such that for any point (x, y, z) in W_1, with $z \neq 0$, the point $(x, y, -z)$ has a $\cos\phi$ value with the same magnitude but opposite sign of the $\cos\phi$ value for (x, y, z). Furthermore, if $z = 0$, then (x, y, z) has a $\cos\phi$ value of 0. Thus, with $\cos\phi$ positive on the top half of the sphere and negative on the bottom half, the integral will cancel out and be equal to zero.

21. **(a)** In cylindrical coordinates, the cone is $z = r$ and the sphere is $r^2 + z^2 = 4$. The surfaces intersect where $z^2 + z^2 = 2z^2 = 4$. So $z = \sqrt{2}$ and $r = \sqrt{2}$.

$$\text{Volume} = \int_0^{2\pi}\int_0^{\sqrt{2}}\int_r^{\sqrt{4-r^2}} r\,dz dr d\theta.$$

(b) In spherical coordinates, the cone is $\phi = \pi/4$ and the sphere is $\rho = 2$.

$$\text{Volume} = \int_0^{2\pi}\int_0^{\pi/4}\int_0^2 \rho^2 \sin\phi \, d\rho d\phi d\theta.$$

25. **(a)** In Cartesian coordinates, the bottom half of the sphere $x^2 + y^2 + z^2 = 1$ is given by $z = -\sqrt{1 - x^2 - y^2}$. Thus

$$\int_W dV = \int_0^1\int_0^{\sqrt{1-x^2}}\int_{-\sqrt{1-x^2-y^2}}^0 dz dy dx.$$

(b) In cylindrical coordinates, the sphere is $r^2 + z^2 = 1$ and the bottom half is given by $z = -\sqrt{1 - r^2}$. Thus

$$\int_W dV = \int_0^{\pi/2}\int_0^1\int_{-\sqrt{1-r^2}}^0 r\,dz dr d\theta.$$

(c) In spherical coordinates, the sphere is $\rho = 1$. Thus,

$$\int_W dV = \int_0^{\pi/2}\int_{\pi/2}^{\pi}\int_0^1 \rho^2 \sin\phi \, d\rho d\phi d\theta.$$

29. Using spherical coordinates:

$$\begin{aligned}
M &= \int_0^{\pi}\int_0^{2\pi}\int_0^3 (3-\rho)\rho^2 \sin\phi \, d\rho \, d\theta \, d\phi \\
&= \int_0^{\pi}\int_0^{2\pi} \left[\rho^3 - \frac{\rho^4}{4}\right]_0^3 \sin\phi \, d\theta \, d\phi \\
&= \frac{27}{4}\int_0^{\pi}\int_0^{2\pi} \sin\phi \, d\theta \, d\phi \\
&= \frac{27}{4}\cdot 2\pi \cdot (-\cos\phi)\bigg|_0^{\pi} = \frac{27}{2}\pi\cdot[-(-1)+1] = 27\pi.
\end{aligned}$$

33. (a) We use the axes shown in Figure 16.26. Then the sphere is given by $r^2+z^2=25$, so

$$\text{Volume} = \int_0^{2\pi}\int_1^5\int_{-\sqrt{25-r^2}}^{\sqrt{25-r^2}} r \, dz dr d\theta.$$

(b) Evaluating gives

$$\begin{aligned}
\text{Volume} = 2\pi\int_1^5 rz\bigg|_{z=-\sqrt{25-r^2}}^{z=\sqrt{25-r^2}} dr &= 2\pi\int_1^5 2r\sqrt{25-r^2}\, dr \\
&= 2\pi\left(-\frac{2}{3}\right)(25-r^2)^{3/2}\bigg|_1^5 \\
&= \frac{4\pi}{3}(24)^{3/2} = 64\sqrt{6}\pi = 492.5 \text{ mm}^3.
\end{aligned}$$

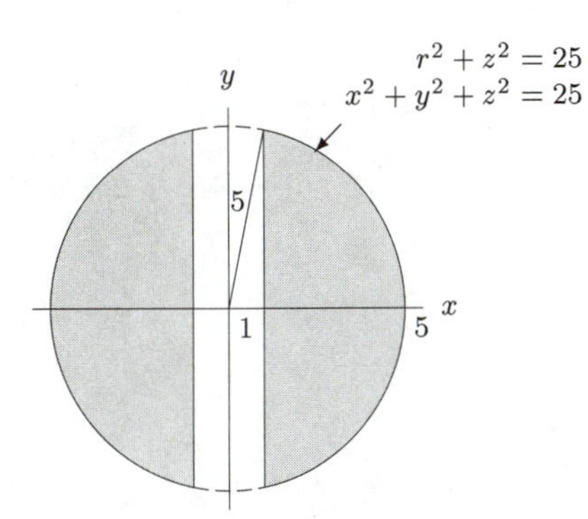

Figure 16.26

37. We first need to find the mass of the solid, using cylindrical coordinates:

$$\begin{aligned}
m &= \int_0^{2\pi}\int_0^1\int_0^{\sqrt{z/a}} r \, dr \, dz \, d\theta \\
&= \int_0^{2\pi}\int_0^1 \frac{z}{2a}\, dz \, d\theta \\
&= \int_0^{2\pi} \frac{1}{4a}\, d\theta = \frac{\pi}{2a}
\end{aligned}$$

It makes sense that the mass would vary inversely with a, since increasing a makes the paraboloid skinnier. Now for the z-coordinate of the center of mass, again using cylindrical coordinates:

$$\begin{aligned}\bar{z} &= \frac{2a}{\pi}\int_0^{2\pi}\int_0^1\int_0^{\sqrt{z/a}} zr\,dr\,dz\,d\theta \\ &= \frac{2a}{\pi}\int_0^{2\pi}\int_0^1 \frac{z^2}{2a}\,dz\,d\theta \\ &= \frac{2a}{\pi}\int_0^{2\pi}\frac{1}{6a}\,d\theta = \frac{2}{3}\end{aligned}$$

41. Assume the base of the cylinder sits on the xy-plane with center at the origin. Because the cylinder is symmetric about the z-axis, the force in the horizontal x or y direction is 0. Thus we need only compute the vertical z component of the force. We are going to use cylindrical coordinates; since the force is $G\cdot\text{mass}/(\text{distance})^2$, a piece of the cylinder of volume dV located at (r,θ,z) exerts on the unit mass a force with magnitude $G(\delta\,dV)/(r^2+z^2)$. See Figure 16.27.

$$\begin{array}{c}\text{Vertical component}\\ \text{of force}\end{array} = \frac{G(\delta\,dV)}{r^2+z^2}\cdot\cos\phi = \frac{G\delta\,dV}{r^2+z^2}\cdot\frac{z}{\sqrt{r^2+z^2}} = \frac{G\delta z\,dV}{(r^2+z^2)^{3/2}}.$$

Adding up all the contributions of all the dV's, we obtain

$$\begin{aligned}\text{Vertical force} &= \int_0^H\int_0^{2\pi}\int_0^R \frac{G\delta zr}{(r^2+z^2)^{3/2}}\,drd\theta dz \\ &= \int_0^H\int_0^{2\pi}(G\delta z)\left(-\frac{1}{\sqrt{r^2+z^2}}\right)\Bigg|_0^R d\theta dz \\ &= \int_0^H\int_0^{2\pi}(G\delta z)\cdot\left(-\frac{1}{\sqrt{R^2+z^2}}+\frac{1}{z}\right)d\theta dz \\ &= \int_0^H 2\pi G\delta\left(1-\frac{z}{\sqrt{R^2+z^2}}\right)dz \\ &= 2\pi G\delta(z-\sqrt{R^2+z^2})\Big|_0^H \\ &= 2\pi G\delta(H-\sqrt{R^2+H^2}+R) = 2\pi G\delta(H+R-\sqrt{R^2+H^2})\end{aligned}$$

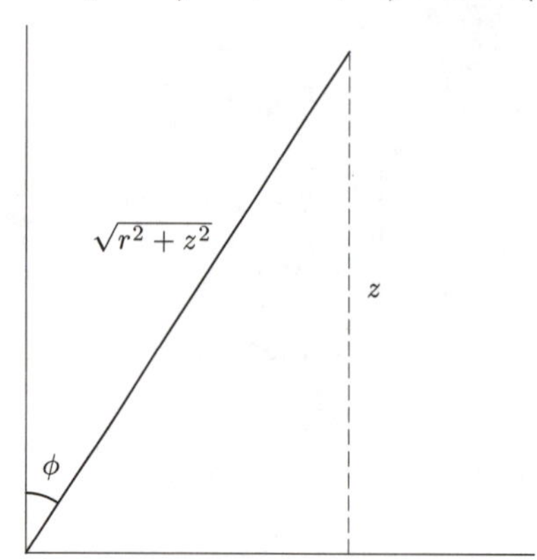

Figure 16.27

45. Using spherical coordinates,

$$\text{Stored energy} = \frac{1}{2}\int_a^b\int_0^{\pi}\int_0^{2\pi}\epsilon E^2\rho^2\sin\phi\,d\theta\,d\phi\,d\rho = \frac{q^2}{32\pi^2\epsilon}\int_a^b\int_0^{\pi}\int_0^{2\pi}\frac{1}{\rho^2}\sin\phi\,d\theta\,d\phi\,d\rho$$
$$= \frac{q^2}{8\pi\epsilon}\int_a^b\frac{1}{\rho^2}\,d\rho = \frac{q^2}{8\pi\epsilon}\left(\frac{1}{a}-\frac{1}{b}\right).$$

Solutions for Section 16.6

Exercises

1. No, p is not a joint density function. Since $p(x, y) = 0$ outside the region R, the volume under the graph of p is the same as the volume under the graph of p over the region R, which is 2 not 1.

5. Yes, p is a joint density function. In the region R we have $1 \geq x^2 + y^2$, so $p(x, y) = (2/\pi)(1 - x^2 - y^2) \geq 0$ for all x and y in R, and $p(x, y) = 0$ for all other (x, y). To check that p is a joint density function, we check that the total volume under the graph of p over the region R is 1. Using polar coordinates, we get:

$$\int_R p(x,y)dA = \frac{2}{\pi}\int_0^{2\pi}\int_0^1 (1-r^2)r\,dr\,d\theta = \frac{2}{\pi}\int_0^{2\pi}\left(\frac{r^2}{2}-\frac{r^4}{4}\right)\Big|_0^1 d\theta = \frac{2}{\pi}\int_0^{2\pi}\frac{1}{4}\,d\theta = 1.$$

9. Since $x + y \leq 3$ for all points (x, y) in the region R, the fraction of the population satisfying $x + y \leq 3$ is 1.

13. The fraction of the population is given by the double integral:

$$\int_0^{1/2}\int_0^1 xy\,dx\,dy = \int_0^{1/2}\frac{x^2y}{2}\Big|_0^1 dy = \int_0^{1/2}\frac{y}{2}dy = \frac{y^2}{4}\Big|_0^{1/2} = \frac{1}{16}.$$

Problems

17. (a) We know that $\int_{-\infty}^{\infty}\int_{-\infty}^{\infty} f(x,y)dydx = 1$ for a joint density function. So,

$$1 = \int_{-\infty}^{\infty}\int_{-\infty}^{\infty} f(x,y)dydx = \int_0^1\int_x^1 kxydydx$$
$$= \frac{1}{8}k$$

hence $k = 8$.

(b) The region where $x < y < \sqrt{x}$ is sketched in Figure 16.28

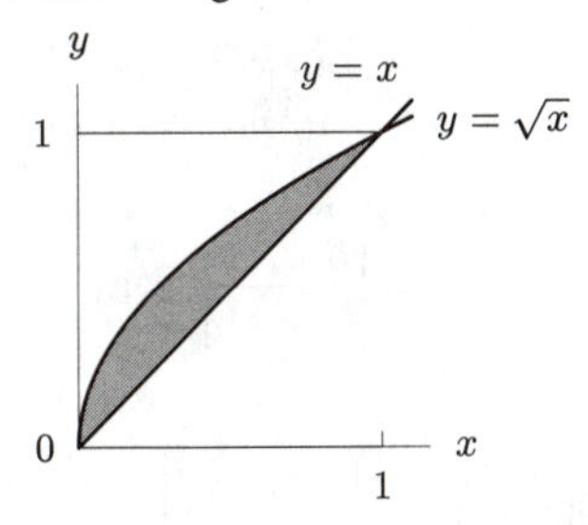

Figure 16.28

So the probability that (x, y) satisfies $x < y < \sqrt{x}$ is given by:

$$\int_0^1\int_x^{\sqrt{x}} 8xy\,dy\,dx = \int_0^1 4x(y^2)\Big|_x^{\sqrt{x}} dx$$
$$= \int_0^1 4x(x - x^2)dx$$
$$= 4\left(\frac{1}{3}x^3 - \frac{1}{4}x^4\right)\Big|_0^1$$
$$= 4\left(\frac{1}{3} - \frac{1}{4}\right)$$
$$= \frac{1}{3}$$

This tells us that in choosing points from the region defined by $0 \leq x \leq y \leq 1$, that $1/3$ of the time we would pick a point from the region defined by $x < y < \sqrt{x}$. These regions are shown in Figure 16.28.

21. (a) Since the exponential function is always positive and λ is positive, $p(t) \geq 0$ for all t, and

$$\int_0^\infty p(t)dt = \lim_{b\to\infty} -e^{-\lambda t}\Big|_0^b = \lim_{b\to\infty} -e^{-bt} + 1 = 1.$$

(b) The density function for the probability that the first substance decays at time t and the second decays at time s is

$$p(t,s) = \lambda e^{-\lambda t}\mu e^{-\mu s} = \lambda\mu e^{-\lambda t-\mu s},$$

for $s \geq 0$ and $t \geq 0$, and is zero otherwise.

(c) We want the probability that the decay time t of the first substance is less than or equal to the decay time s of the second, so we want to integrate the density function over the region $0 \leq t \leq s$. Thus, we compute

$$\begin{aligned}
\int_0^\infty \int_t^\infty \lambda\mu e^{-\lambda t} e^{-\mu s}\, ds\, dt &= \int_0^\infty \lambda e^{-\lambda t}(-e^{-\mu s})\Big|_t^\infty dt \\
&= \int_0^\infty \lambda e^{-\lambda t} e^{-\mu t}\, dt \\
&= \int_0^\infty \lambda e^{(-\lambda+\mu)t}\, dt \\
&= \frac{-\lambda}{\lambda+\mu} e^{-(\lambda+\mu)t}\Big|_0^\infty = \frac{\lambda}{\lambda+\mu}.
\end{aligned}$$

So for example, if $\lambda = 1$ and $\mu = 4$, then the probability that the first substance decays first is $1/5$.

Solutions for Section 16.7

Exercises

1. We have

$$\frac{\partial(x,y)}{\partial(s,t)} = \begin{vmatrix} x_s & x_t \\ y_s & y_t \end{vmatrix} = \begin{vmatrix} 5 & 2 \\ 3 & 1 \end{vmatrix} = -1.$$

Therefore,

$$\left|\frac{\partial(x,y)}{\partial(s,t)}\right| = 1.$$

5. We have

$$\frac{\partial(x,y,z)}{\partial(s,t,u)} = \begin{vmatrix} x_s & x_t & x_u \\ y_s & y_t & y_u \\ z_s & z_t & z_u \end{vmatrix} = \begin{vmatrix} 3 & 1 & 2 \\ 1 & 5 & -1 \\ 2 & -1 & 1 \end{vmatrix}.$$

This 3×3 determinant is computed the same way as for the cross product, with the entries 3, 1, 2 in the first row playing the same role as $\vec{i}, \vec{j}, \vec{k}$. We get

$$\frac{\partial(x,y,z)}{\partial(s,t,u)} = ((5)(1) - (-1)(-1))3 + ((-1)(2) - (1)(1))1 + ((1)(-1) - (2)(5))2 = -13.$$

Problems

9. Given

$$\begin{cases} x = \rho\sin\phi\cos\theta \\ y = \rho\sin\phi\sin\theta \\ z = \rho\cos\phi, \end{cases}$$

$$\frac{\partial(x,y,z)}{\partial(\rho,\phi,\theta)} = \begin{vmatrix} \frac{\partial x}{\partial \rho} & \frac{\partial x}{\partial \phi} & \frac{\partial x}{\partial \theta} \\ \frac{\partial y}{\partial \rho} & \frac{\partial y}{\partial \phi} & \frac{\partial y}{\partial \theta} \\ \frac{\partial z}{\partial \rho} & \frac{\partial z}{\partial \phi} & \frac{\partial z}{\partial \theta} \end{vmatrix} = \begin{vmatrix} \sin\phi\cos\theta & \rho\cos\phi\cos\theta & -\rho\sin\phi\sin\theta \\ \sin\phi\sin\theta & \rho\cos\phi\sin\theta & \rho\sin\phi\cos\theta \\ \cos\phi & -\rho\sin\phi & 0 \end{vmatrix}$$

$$= \cos\phi \begin{vmatrix} \rho\cos\phi\cos\theta & -\rho\sin\phi\sin\theta \\ \rho\cos\phi\sin\theta & \rho\sin\phi\cos\theta \end{vmatrix} + \rho\sin\phi \begin{vmatrix} \sin\phi\cos\theta & -\rho\sin\phi\sin\theta \\ \sin\phi\sin\theta & \rho\sin\phi\cos\theta \end{vmatrix}$$

$$= \cos\phi(\rho^2\cos^2\theta\cos\phi\sin\phi + \rho^2\sin^2\theta\cos\phi\sin\phi)$$

$$+\rho\sin\phi(\rho\sin^2\phi\cos^2\theta + \rho\sin^2\phi\sin^2\theta)$$

$$= \rho^2\cos^2\phi\sin\phi + \rho^2\sin^3\phi$$

$$= \rho^2\sin\phi.$$

13. Given

$$\begin{cases} s = xy \\ t = xy^2, \end{cases}$$

we have

$$\frac{\partial(s,t)}{\partial(x,y)} = \begin{vmatrix} \frac{\partial s}{\partial x} & \frac{\partial s}{\partial y} \\ \frac{\partial t}{\partial x} & \frac{\partial t}{\partial y} \end{vmatrix} = \begin{vmatrix} y & x \\ y^2 & 2xy \end{vmatrix} = xy^2 = t.$$

Since

$$\frac{\partial(s,t)}{\partial(x,y)} \cdot \frac{\partial(x,y)}{\partial(s,t)} = 1,$$

$$\frac{\partial(x,y)}{\partial(s,t)} = t \qquad \text{so} \qquad \left|\frac{\partial(x,y)}{\partial(s,t)}\right| = \frac{1}{t}$$

So

$$\int_R xy^2\,dA = \int_T t\left|\frac{\partial(x,y)}{\partial(s,t)}\right| ds\,dt = \int_T t(\frac{1}{t})\,ds\,dt = \int_T ds\,dt,$$

where T is the region bounded by $s = 1$, $s = 4$, $t = 1$, $t = 4$.
Then

$$\int_R xy^2\,dA = \int_1^4 ds \int_1^4 dt = 9.$$

Solutions for Chapter 16 Review

Exercises

1.

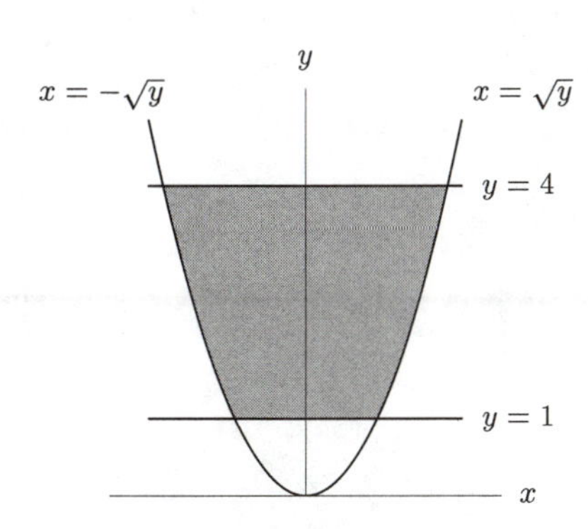

Figure 16.29

5. We use Cartesian coordinates, oriented so that the cube is in the first quadrant. See Figure 16.30. Then, if f is an arbitrary function, the integral is

$$\int_0^2 \int_0^3 \int_0^5 f\,dxdydz.$$

Other answers are possible. In particular, the order of integration can be changed.

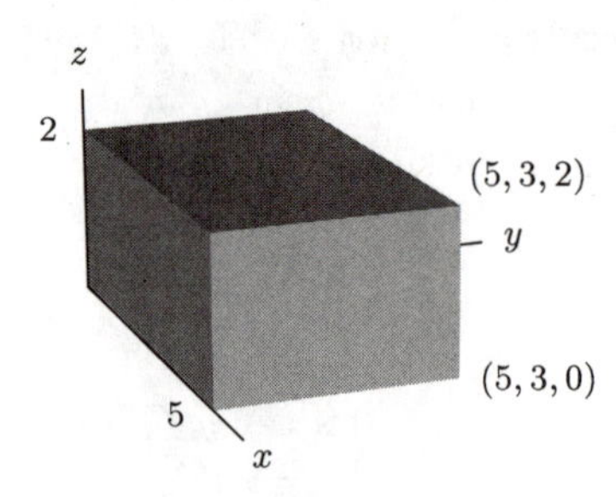

Figure 16.30

9.

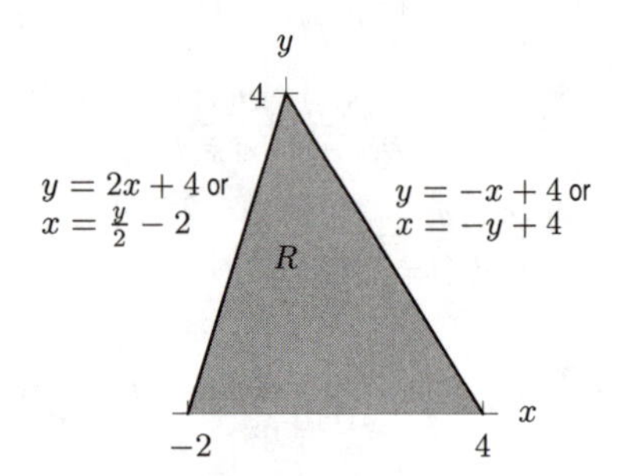

Figure 16.31

Integrating with respect to x first we get

$$\int_0^4 \int_{\frac{y}{2}-2}^{-y+4} f(x, y)\,dx\,dy$$

Integrating with respect to y first we get

$$\int_{-2}^0 \int_0^{2x+4} f(x, y)\,dy\,dx + \int_0^4 \int_0^{-x+4} f(x, y)\,dy\,dx.$$

13.

$$\begin{aligned}
\int_0^1 \int_0^y (\sin^3 x)(\cos x)(\cos y)\,dx\,dy &= \int_0^1 (\cos y)\left[\frac{\sin^4 x}{4}\right]\Bigg|_0^y dy \\
&= \frac{1}{4}\int_0^1 (\sin^4 y)(\cos y)\,dy \\
&= \frac{\sin^5 y}{20}\Bigg|_0^1 \\
&= \frac{\sin^5 1}{20}.
\end{aligned}$$

17. From Figure 16.32, we have the following iterated integrals:

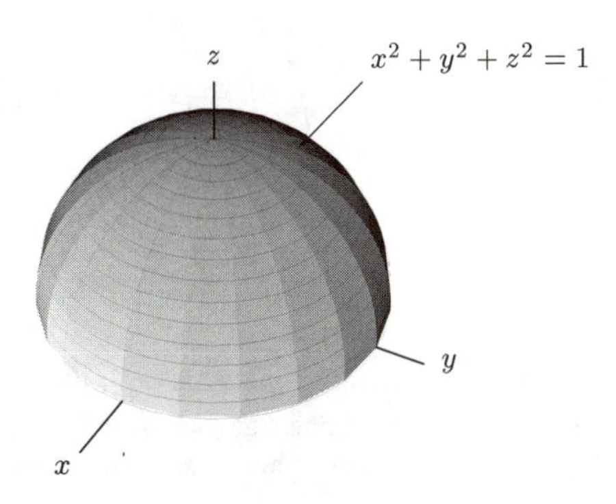

Figure 16.32

(a) $\int_R f\,dV = \int_{-1}^{1}\int_{-\sqrt{1-x^2}}^{\sqrt{1-x^2}}\int_0^{\sqrt{1-x^2-y^2}} f(x,y,z)\,dzdydx$

(b) $\int_R f\,dV = \int_{-1}^{1}\int_{-\sqrt{1-y^2}}^{\sqrt{1-y^2}}\int_0^{\sqrt{1-x^2-y^2}} f(x,y,z)\,dzdxdy$

(c) $\int_R f\,dV = \int_{-1}^{1}\int_0^{\sqrt{1-y^2}}\int_{-\sqrt{1-y^2-z^2}}^{\sqrt{1-y^2-z^2}} f(x,y,z)\,dxdzdy$

(d) $\int_R f\,dV = \int_{-1}^{1}\int_0^{\sqrt{1-x^2}}\int_{-\sqrt{1-x^2-z^2}}^{\sqrt{1-x^2-z^2}} f(x,y,z)\,dydzdx$

(e) $\int_R f\,dV = \int_0^{1}\int_{-\sqrt{1-z^2}}^{\sqrt{1-z^2}}\int_{-\sqrt{1-x^2-z^2}}^{\sqrt{1-x^2-z^2}} f(x,y,z)\,dydxdz$

(f) $\int_R f\,dV = \int_0^{1}\int_{-\sqrt{1-z^2}}^{\sqrt{1-z^2}}\int_{-\sqrt{1-y^2-z^2}}^{\sqrt{1-y^2-z^2}} f(x,y,z)\,dxdydz$

21. W is a cylindrical shell, so cylindrical coordinates should be used. See Figure 16.33.

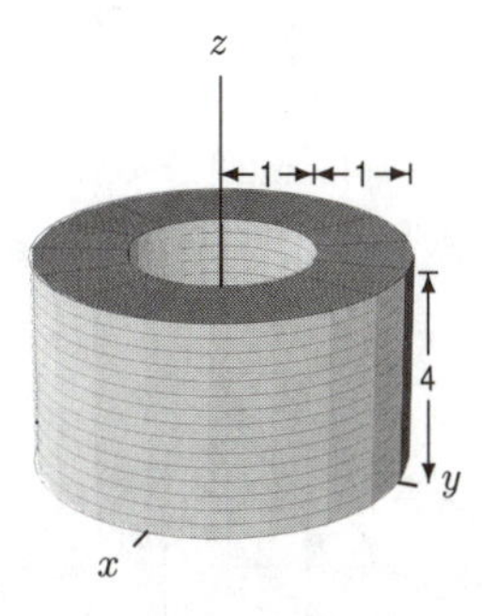

Figure 16.33

$$\begin{aligned}
\int_W \frac{z}{(x^2+y^2)^{3/2}}\,dV &= \int_0^4\int_0^{2\pi}\int_1^2 \frac{z}{r^3}\,rdr\,d\theta\,dz \\
&= \int_0^4\int_0^{2\pi}\int_1^2 \frac{z}{r^2}\,dr\,d\theta\,dz \\
&= \int_0^4\int_0^{2\pi} \left(-\frac{z}{r}\right)\Big|_1^2\,d\theta\,dz
\end{aligned}$$

$$= \int_0^4 \int_0^{2\pi} \frac{z}{2}\, d\theta\, dz$$
$$= \int_0^4 \frac{z}{2} \cdot 2\pi\, dz = \frac{1}{2}\pi \cdot z^2 \bigg|_0^4 = 8\pi$$

Problems

25. Can't tell, since y^3 is both positive and negative for $x < 0$.

29. Zero. You can see this in several ways. One way is to observe that xy is positive on the part of the sphere above and below the first and third quadrants (where x and y are of the same sign) and negative (of equal absolute value) on the part of the sphere above and below the second and fourth quadrants (where x and y have opposite signs). These add up to zero in the integral of xy over all of W.

Another way to see that the integral is zero is to write the triple integral as an iterated integral, say integrating first with respect to x. For fixed y and z, the x-integral is over an interval symmetric about 0. The integral of x over such an interval is zero. If any of the inner integrals in an iterated integral is zero, then the triple integral is zero.

33. Negative. Since $z^2 - 1 \le 0$ in the sphere, its integral is negative.

37. Let the lower left part of the forest be at $(0, 0)$. Then the other corners have coordinates as shown. The population density function is then given by

$$\rho(x, y) = 10 - 2y$$

The equations of the two diagonal lines are $x = -2y/5$ and $x = 6 + 2y/5$. So the total rabbit population in the forest is

$$\int_0^5 \int_{-\frac{2}{5}y}^{6+\frac{2}{5}y} (10 - 2y)\, dx\, dy = \int_0^5 (10 - 2y)(6 + \frac{4}{5}y)\, dy$$
$$= \int_0^5 (60 - 4y - \frac{8}{5}y^2)\, dy$$
$$= (60y - 2y^2 - \frac{8}{15}y^3)\bigg|_0^5$$
$$= 300 - 50 - \frac{8}{15} \cdot 125$$
$$= \frac{2750}{15} = \frac{550}{3} \approx 183$$

41. Since the hole resembles a cylinder, we will use cylindrical coordinates. Let the center of the sphere be at the origin, and let the center of the hole be the z-axis (see Figure 16.34).

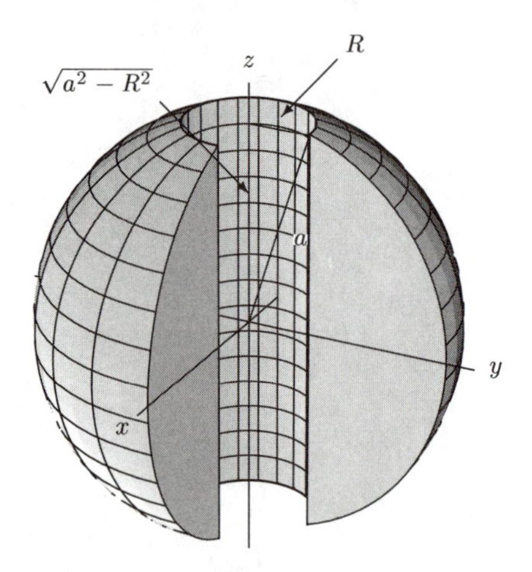

Figure 16.34

Then we will integrate from $z = -\sqrt{a^2 - R^2}$ to $z = \sqrt{a^2 - R^2}$, and each cross-section will be an annulus. So the volume is

$$\begin{aligned}
\int_{-\sqrt{a^2-R^2}}^{\sqrt{a^2-R^2}} \int_0^{2\pi} \int_R^{\sqrt{a^2-z^2}} r\, dr\, d\theta\, dz &= \int_{-\sqrt{a^2-R^2}}^{\sqrt{a^2-R^2}} \int_0^{2\pi} \frac{1}{2}(a^2 - z^2 - R^2)\, d\theta\, dz \\
&= \pi \int_{-\sqrt{a^2-R^2}}^{\sqrt{a^2-R^2}} (a^2 - z^2 - R^2)\, dz \\
&= \pi \left[(a^2 - R^2)(2\sqrt{a^2 - R^2}) - \frac{1}{3}(2(a^2 - R^2)^{\frac{3}{2}}) \right] \\
&= \frac{4\pi}{3}(a^2 - R^2)^{\frac{3}{2}}
\end{aligned}$$

45. Let the ball be centered at the origin. Since a ball looks the same from all directions, we can choose the axis of rotation; in this case, let it be the z-axis. It is best to use spherical coordinates, so then

$$\begin{aligned}
x^2 + y^2 &= (\rho \sin\phi \cos\theta)^2 + (\rho \sin\phi \sin\theta)^2 \\
&= \rho^2 \sin^2\phi
\end{aligned}$$

Then $m/v = \text{Density} = 1$, so the moment of inertia is

$$\begin{aligned}
I_z &= \int_0^R \int_0^{2\pi} \int_0^{\pi} 1(\rho^2 \sin^2\phi)\rho^2 \sin\phi\, d\phi\, d\theta\, d\rho \\
&= \int_0^R \int_0^{2\pi} \int_0^{\pi} \rho^4 (\sin\phi)(1 - \cos^2\phi)\, d\phi\, d\theta\, d\rho \\
&= \int_0^R \int_0^{2\pi} \rho^4 \left(-\cos\phi + \frac{1}{3}\cos^3\phi\right)\Big|_0^{\pi} d\theta\, d\rho \\
&= \int_0^R \int_0^{2\pi} \frac{4}{3}\rho^4\, d\theta\, d\rho \\
&= \int_0^R \frac{8\pi}{3}\rho^4\, d\rho = \frac{8}{15}\pi R^5
\end{aligned}$$

CAS Challenge Problems

49. The region is the triangle to the right of the y-axis, below the line $y = 1$, and above the line $y = x$. Thus the integral can be written as $\int_0^1 \int_x^1 e^{y^2}\, dydx$ or as $\int_0^1 \int_0^y e^{y^2}\, dxdy$. The second of these integrals can be evaluated easily by hand:

$$\begin{aligned}
\int_0^1 \int_0^y e^{y^2}\, dxdy &= \int_0^1 \left(e^{y^2} x \Big|_{x=0}^{x=y} \right) dy = \int_0^1 y e^{y^2}\, dy \\
&= \frac{1}{2} e^{y^2} \Big|_0^1 = \frac{1}{2}(e - 1)
\end{aligned}$$

The other integral cannot be done by hand with the methods you have learned, but some computer algebra systems will compute it and give the same answer.

CHECK YOUR UNDERSTANDING

1. False. For example, if $f(x, y) < 0$ for all (x, y) in the region R, then $\int_R f\, dA$ is negative.

5. True. The double integral is the limit of the sum $\sum_{\Delta A \to 0} \rho(x,y)\Delta A$. Each of the terms $\rho(x,y)\Delta A$ is an approximation of the total population inside a small rectangle of area ΔA. Thus the limit of the sum of all of these numbers as $\Delta A \to \infty$ gives the total population of the region R.

9. False. There is no reason to expect this to be true, since the behavior of f on one half of R can be completely unrelated to the behavior of f on the other half. As a counterexample, suppose that f is defined so that $f(x,y) = 0$ for points (x,y) lying in S, and $f(x,y) = 1$ for points (x,y) lying in the part of R that is not in S. Then $\int_S f\,dA = 0$, since $f = 0$ on all of S. To evaluate $\int_R f\,dA$, note that $f = 1$ on the square S_1 which is $0 \le x \le 1, 1 \le y \le 2$. Then $\int_R f\,dA = \int_{S_1} f\,dA = \text{Area}(S_1) = 1$, since $f = 0$ on S.

13. True. For any point in the region of integration we have $1 \le x \le 2$, and so y is between the positive numbers 1 and 8.

17. False. The given limits describe only the upper half disk where $y \ge 0$. The correct limits are $\int_{-a}^{a} \int_{-\sqrt{a^2-x^2}}^{\sqrt{a^2-x^2}} f dydx$.

21. False. The integral gives the total mass of the material contained in W.

25. True. Both sets of limits describe the solid region lying above the rectangle $-1 \le x \le 1, 0 \le y \le 1, z = 0$ and below the parabolic cylinder $z = 1 - x^2$.

29. False. As a counterexample, let W_1 be the solid cube $0 \le x \le 1, 0 \le y \le 1, 0 \le z \le 1$, and let W_2 be the solid cube $-\frac{1}{2} \le x \le 0, -\frac{1}{2} \le y \le 0, -\frac{1}{2} \le z \le 0$. Then volume$(W_1) = 1$ and volume$(W_2) = \frac{1}{8}$. Now if $f(x,y,z) = -1$, then $\int_{W_1} f\,dV = 1 \cdot -1$ which is less than $\int_{W_2} f\,dV = \frac{1}{8} \cdot -1$.

CHAPTER SEVENTEEN

Solutions for Section 17.1

Exercises

1. The direction vectors of the lines, $-\vec{i}+4\vec{j}-2\vec{k}$ and $2\vec{i}-8\vec{j}+4\vec{k}$, are multiples of each other (the second is -2 times the first). Thus the lines are parallel. To see if they are the same line, we take the point corresponding to $t=0$ on the first line, which has position vector $3\vec{i}+3\vec{j}-\vec{k}$, and see if it is on the second line. So we solve

$$(1+2t)\vec{i}+(11-8t)\vec{j}+(4t-5)\vec{k}=3\vec{i}+3\vec{j}-\vec{k}.$$

This has solution $t=1$, so the two lines have a point in common and must be the same line, parameterized in two different ways.

5. The xz-plane is $y=0$, so one possible answer is

$$x=2\cos t,\quad y=0,\quad z=2\sin t.$$

9. The xy-plane is $z=0$, so a possible answer is

$$x=t,\quad y=t^3,\quad z=0.$$

13. Since its diameters lie along the x and y-axes and its center is the origin, the ellipse must lie in the xy-plane, hence at $z=0$. The x-coordinate ranges between -3 and 3 and the y-coordinate between -2 and 2. One possible answer is

$$x=3\cos t,\quad y=2\sin t,\quad z=0.$$

17. One possible parameterization is

$$x=3+t,\quad y=2t,\quad z=-4-t.$$

21. One possible parameterization is

$$x=1,\quad y=0,\quad z=t.$$

25. The line passes through $(3,0,0)$ and $(0,0,-5)$. The displacement vector from the first of these points to the second is $\vec{v}=-3\vec{i}-5\vec{k}$. The line through point $(3,0,0)$ and with direction vector $\vec{v}=-3\vec{i}-5\vec{k}$ is given by parametric equations

$$x=3-3t,$$
$$y=0,$$
$$z=-5t.$$

Other parameterizations of the same line are also possible.

29. The vector from P_0 to P_1 is $\vec{v}=(5+1)\vec{i}+(2+3)\vec{j}=6\vec{i}+5\vec{j}$. Since $P_0=-\vec{i}-3\vec{j}$, the line is

$$\vec{r}(t)=-\vec{i}-3\vec{j}+t(6\vec{i}+5\vec{j})\quad \text{for } 0\le t\le 1.$$

In coordinate form, the equations are $x=-1+6t$, $y=-3+5t$, $0\le t\le 1$

33. The direction vectors for these two lines are $\vec{v}_1=-2\vec{i}+\vec{j}+2\vec{k}$ and $\vec{v}_2=6\vec{i}-3\vec{j}-6\vec{k}$. Since $\vec{v}_2=-3\vec{v}_1$, the lines have the same direction. To see that the lines are not the same, check that a point on the first line (say $(3,5,-4)$) is not on the second line: if $3=7+6t$, then $t=-2/3$, so $y=1-3(-2/3)=3\neq 5$. So the lines are parallel and do not intersect.

Problems

37. We find the parameterization in terms of the displacement vector $\overrightarrow{OP}=2\vec{i}+5\vec{j}$ from the origin to the point P and the displacement vector $\overrightarrow{PQ}=10\vec{i}+4\vec{j}$ from P to Q.
$\vec{r}(t)=\overrightarrow{OP}+(t/5)\overrightarrow{PQ}$ or $\vec{r}(t)=(2+(t/5)10)\vec{i}+(5+(t/5)4)\vec{j}$

41. The graph is parameterized by $x = t$, $y = \sqrt{t}$. To obtain the segment, we restrict t to $1 \leq t \leq 16$. Thus one possible answer is

$$x = t, \quad y = \sqrt{t}, \qquad 1 \leq t \leq 16.$$

45. It is a straight line through the point $(3, 5, 7)$ parallel to the vector $\vec{i} - \vec{j} + 2\vec{k}$. A linear parameterization of the same line is $x = 3 + t$, $y = 5 - t$, $z = 7 + 2t$.

49. Add the two equations to get $3x = 8$, or $x = \frac{8}{3}$. Then we have

$$-y + z = \frac{1}{3}.$$

So a possible parameterization is

$$x = \frac{8}{3}, \quad y = t, \quad z = \frac{1}{3} + t.$$

53. These equations parameterize a line. Since $(3 + t) + (2t) + 3(1 - t) = 6$, we have $x + y + 3z = 6$. Similarly, $x - y - z = (3 + t) - 2t - (1 - t) = 2$. That is, the curve lies entirely in the plane $x + y + 3z = 6$ and in the plane $x - y - z = 2$. Since the normals to the two planes, $\vec{n_1} = \vec{i} + \vec{j} + 3\vec{k}$ and $\vec{n_2} = \vec{i} - \vec{j} - \vec{k}$ are not parallel, the line is the intersection of two nonparallel planes, which is a straight line in 3-dimensional space.

57. The three shadows appear as a circle, a cosine wave and a sine wave, respectively.

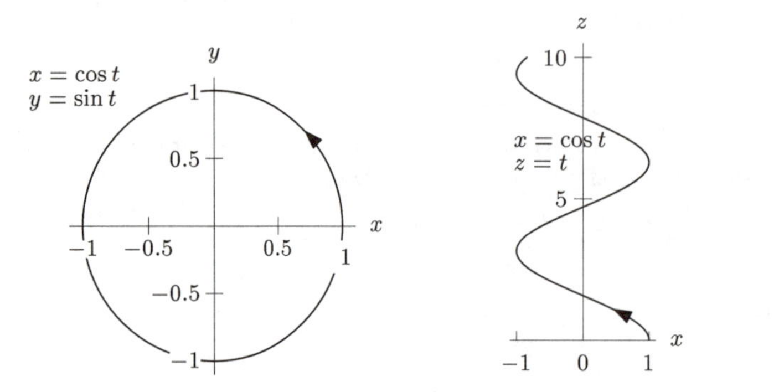

Figure 17.1

61. (a) Parametric equations are

$$x = 1 + 2t, \quad y = 5 + 3t, \quad z = 2 - t.$$

(b) We want to minimize D, the square of the distance of a point to the origin, where

$$D = (x - 0)^2 + (y - 0)^2 + (z - 0)^2 = (1 + 2t)^2 + (5 + 3t)^2 + (2 - t)^2.$$

Differentiating to find the critical points gives

$$\begin{aligned} \frac{dD}{dt} = 2(1 + 2t)2 + 2(5 + 3t)3 + 2(2 - t)(-1) &= 0 \\ 2 + 4t + 15 + 9t - 2 + t &= 0 \\ t &= \frac{-15}{14}. \end{aligned}$$

Thus

$$\begin{aligned} x &= 1 + 2\left(\frac{-15}{14}\right) = \frac{-8}{7} \\ y &= 5 + 3\left(\frac{-15}{14}\right) = \frac{25}{14} \\ z &= 2 - \left(\frac{-15}{14}\right) = \frac{43}{14}. \end{aligned}$$

Since the distance of the point on the line from the origin increases without bound as the magnitude of x, y, z increase, the only critical point of D must be a global minimum. Therefore, the point $(-8/7, 25/14, 43/14)$ is the point on the line closest to the origin.

Solutions for Section 17.2

Exercises

1. To find $\vec{v}(t)$ we first find $dx/dt = 6t$ and $dy/dt = 3t^2$. Therefore, the velocity vector is $\vec{v} = 6t\vec{i} + 3t^2\vec{j}$. The speed of the particle is given by the magnitude of the vector,

$$\|\vec{v}\| = \sqrt{\left(\frac{dx}{dt}\right)^2 + \left(\frac{dy}{dt}\right)^2} = \sqrt{(6t)^2 + (3t^2)^2} = 3|t| \cdot \sqrt{4+t^2}.$$

The particle stops when $\vec{v} = \vec{0}$, so when $6t = 3t^2 = 0$. Therefore, the particle stops when $t = 0$.

5. The velocity vector $\vec{v}$ is given by:

$$\vec{v} = \frac{d}{dt}(3\cos t)\vec{i} + \frac{d}{dt}(4\sin t)\vec{j} = -3\sin t\vec{i} + 4\cos t\vec{j}.$$

The acceleration vector $\vec{a}$ is given by:

$$\vec{a} = \frac{d\vec{v}}{dt} = \frac{d}{dt}(-3\sin t)\vec{i} + \frac{d}{dt}(4\cos t)\vec{j} = -3\cos t\vec{i} - 4\sin t\vec{j}.$$

9. At $t = 3$, the particle is at the point $(4, 5)$ since $x = 3^3 - 8\cdot 3 + 1 = 4$ and $y = 3^2 - 4 = 5$. The velocity vector is

$$\vec{v} = \frac{dx}{dt}\vec{i} + \frac{dy}{dt}\vec{j} = (3t^2 - 8)\vec{i} + (2t)\vec{j}.$$

When $t = 3$, the velocity vector is $\vec{v} = 19\vec{i} + 6\vec{j}$. The speed of the particle at $t = 3$ is given by

$$\text{Speed} = \|\vec{v}\| = \sqrt{19^2 + 6^2} = \sqrt{397} = 19.92.$$

13. In vector form the parameterization is

$$\vec{r} = 2\vec{i} + 3\vec{j} + 5\vec{k} + t^2(\vec{i} - 2\vec{j} - \vec{k}).$$

Thus the motion is along the straight line through $(2, 3, 5)$ in the direction of $\vec{i} - 2\vec{j} - \vec{k}$. The velocity vector $\vec{v}$ is

$$\vec{v} = \frac{dx}{dt}\vec{i} + \frac{dy}{dt}\vec{j} + \frac{dz}{dt}\vec{k} = 2t(\vec{i} - 2\vec{j} - \vec{k})$$

The acceleration vector $\vec{a}$ is

$$\vec{a} = \frac{d^2x}{dt^2}\vec{i} + \frac{d^2y}{dt^2}\vec{j} + \frac{d^2z}{dt^2}\vec{k} = 2(\vec{i} - 2\vec{j} - \vec{k}).$$

The speed is

$$\|\vec{v}\| = 2|t|\|\vec{i} - 2\vec{j} - \vec{k}\| = 2\sqrt{6}|t|.$$

The acceleration vector is constant and points in the direction of $\vec{i} - 2\vec{j} - \vec{k}$. When $t < 0$ the absolute value $|t|$ is decreasing, hence the speed is decreasing. Also, when $t < 0$ the velocity vector $2t(\vec{i} - 2\vec{j} - \vec{k})$ points in the direction opposite to $\vec{i} - 2\vec{j} - \vec{k}$. When $t > 0$ the absolute value $|t|$ is increasing and hence the speed is increasing. Also, when $t > 0$ the velocity vector points in the same direction as $\vec{i} - 2\vec{j} - \vec{k}$.

17. We have

$$\text{Length} = \int_0^{2\pi} \sqrt{(-3\sin 3t)^2 + (5\cos 5t)^2}\, dt.$$

We cannot find this integral symbolically, but numerical methods show Length ≈ 24.6.

Problems

21. At $t = 2$, the position and velocity vectors are

$$\vec{r}(2) = (2-1)^2\vec{i} + 2\vec{j} + (2 \cdot 2^3 - 3 \cdot 2^2)\vec{k} = \vec{i} + 2\vec{j} + 4\vec{k},$$
$$\vec{v}(2) = 2 \cdot (2-1)\vec{i} + (6 \cdot 2^2 - 6 \cdot 2)\vec{k} = 2\vec{i} + 12\vec{k}.$$

So we want the line going through the point $(1, 2, 4)$ at the time $t = 2$, in the direction $2\vec{i} + 12\vec{k}$:

$$x = 1 + 2(t-2), \quad y = 2 \quad z = 4 + 12(t-2).$$

25. At time t the particle is $s = t - 7$ seconds from P, so the displacement vector from the point P to the particle is $\vec{d} = s\vec{v}$. To find the position vector of the particle at time t, we add this to the position vector $\vec{r}_0 = 5\vec{i} + 4\vec{j} + 3\vec{k}$ for the point P. Thus a vector equation for the motion is:

$$\begin{aligned}\vec{r} &= \vec{r}_0 + s\vec{v} \\ &= (5\vec{i} + 4\vec{j} + 3\vec{k}) + (t-7)(3\vec{i} + \vec{j} + 2\vec{k}),\end{aligned}$$

or equivalently,

$$x = 5 + 3(t-7), \quad y = 4 + 1(t-7), \quad z = 3 + 2(t-7).$$

Notice that these equations are linear. They describe motion on a straight line through the point $(5, 4, 3)$ that is parallel to the velocity vector $\vec{v} = 3\vec{i} + \vec{j} + 2\vec{k}$.

29. Parametric equations for a line in 2-space are

$$\begin{aligned}x &= x_0 + at \\ y &= y_0 + bt\end{aligned}$$

where (x_0, y_0) is a point on the line and $\vec{v} = a\vec{i} + b\vec{j}$ is the direction of motion. Notice that the slope of the line is equal to $\Delta y/\Delta x = b/a$, so in this case we have

$$\begin{aligned}\frac{b}{a} &= \text{Slope} = -2, \\ b &= -2a.\end{aligned}$$

In addition, the speed is 3, so we have

$$\begin{aligned}\|\vec{v}\| &= 3 \\ \sqrt{a^2 + b^2} &= 3 \\ a^2 + b^2 &= 9.\end{aligned}$$

Substituting $b = -2a$ gives

$$\begin{aligned}a^2 + (-2a)^2 &= 9 \\ 5a^2 &= 9 \\ a &= \frac{3}{\sqrt{5}}, -\frac{3}{\sqrt{5}}.\end{aligned}$$

If we use $a = 3/\sqrt{5}$, then $b = -2a = -6/\sqrt{5}$. The point (x_0, y_0) can be any point on the line: we use $(0, 5)$. The parametric equations are

$$x = \frac{3}{\sqrt{5}}t, \quad y = 5 - \frac{6}{\sqrt{5}}t.$$

Alternatively, we can use $a = -3/\sqrt{5}$ giving $b = 6/\sqrt{5}$. An alternative answer, which represents the particle moving in the opposite direction is

$$x = -\frac{3}{\sqrt{5}}t, \quad y = 5 + \frac{6}{\sqrt{5}}t.$$

33. **(a)** No. The height of the particle is given by $2t$; the vertical velocity is the derivative $d(2t)/dt = 2$. Because this is a positive constant, the vertical component of the velocity vector is upward at a constant speed of 2.

(b) When $2t = 10$, so $t = 5$.

(c) The velocity vector is given by

$$\vec{v}(t) = \frac{d\vec{r}}{dt} = \frac{dx}{dt}\vec{i} + \frac{dy}{dt}\vec{j} + \frac{dz}{dt}\vec{k}$$
$$= -(\sin t)\vec{i} + (\cos t)\vec{j} + 2\vec{k}.$$

From (b), the particle is at 10 units above the ground when $t = 5$, so at $t = 5$,

$$\vec{v}(5) = 0.959\vec{i} + 0.284\vec{j} + 2\vec{k}.$$

Therefore, $\vec{v}(5) = -\sin(5)\vec{i} + \cos(5)\vec{j} + 2\vec{k}$.

(d) At this point, $t = 5$, the particle is located at

$$\vec{r}(5) = (\cos(5), \sin(5), 10) = (0.284, -0.959, 10).$$

The tangent vector to the helix at this point is given by the velocity vector found in part (c), that is, $\vec{v}(5) = 0.959\vec{i} + 0.284\vec{j} + 2\vec{k}$. So, the equation of the tangent line is

$$\vec{r}(t) = 0.284\vec{i} - 0.959\vec{j} + 10\vec{k} + (t-5)(0.959\vec{i} + 0.284\vec{j} + 2\vec{k}).$$

37. At time t object B is at the point with position vector $\vec{r}_B(t) = \vec{r}_A(2t)$, which is exactly where object A is at time $2t$. Thus B visits the same points as A, but does so at different times; A gets there later. While B covers the same path as A, it moves twice as fast. To see this, note for example that between $t = 1$ and $t = 3$, object B moves along the path from $\vec{r}_B(1) = \vec{r}_A(2)$ to $\vec{r}_B(3) = \vec{r}_A(6)$ which is traversed by object A during the time interval from $t = 2$ to $t = 6$. It takes A twice as long to cover the same ground.

In the case where $\vec{r}_A(t) = t\vec{i} + t^2\vec{j}$, both objects move on the parabola $y = x^2$. Both A and B are at the origin at time $t = 0$, but B arrives at the point $(2, 4)$ at time $t = 1$, whereas A does does not get there until $t = 2$.

41. The acceleration vector points from the object to the center of the orbit, and the velocity vector points from the object tangent to the circle in the direction of motion. From Figure 17.2 we see that the movement is counterclockwise.

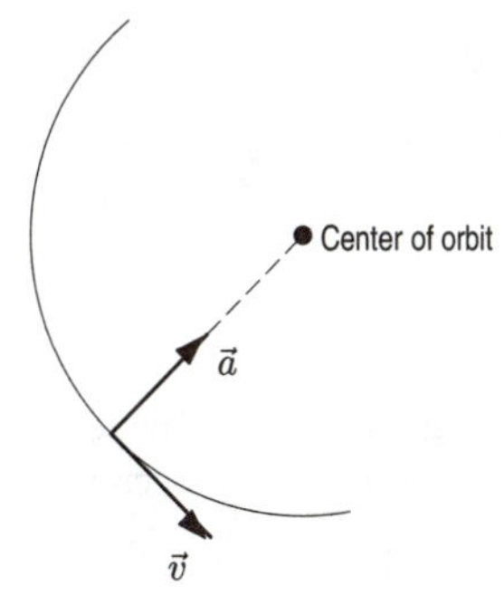

Figure 17.2

Solutions for Section 17.3

Exercises

1. Notice that for a repulsive force, the vectors point outward, away from the particle at the origin, for an attractive force, the vectors point toward the particle. So we can match up the vector field with the description as follows:

(a) IV
(b) III
(c) I
(d) II

5.

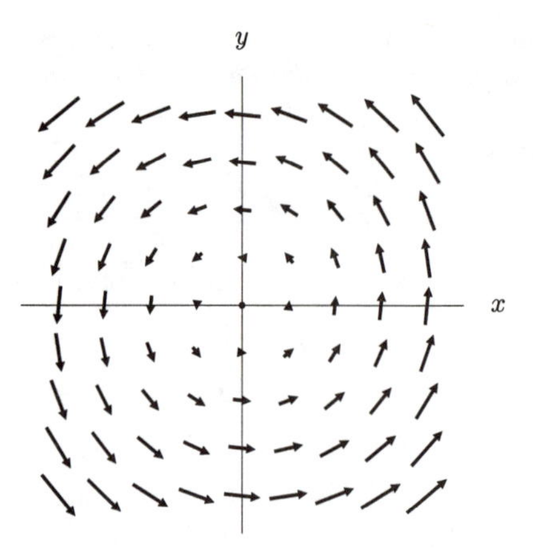

Figure 17.3: $\vec{F}(x, y) = -y\vec{i} + x\vec{j}$

9.

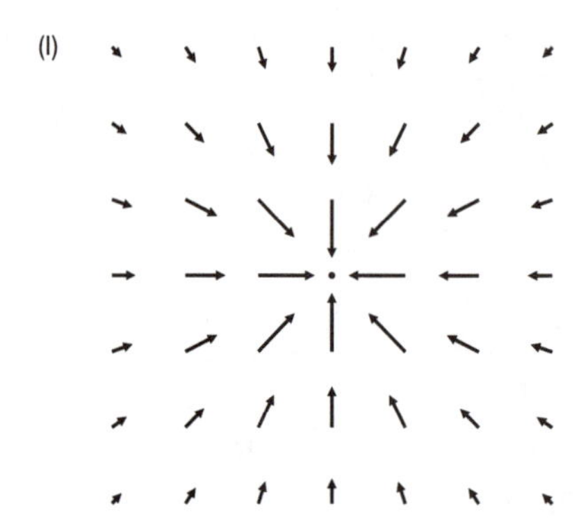

Figure 17.4: $\vec{F}(\vec{r}) = -\vec{r}/\|\vec{r}\|^3$

13. $\vec{V} = x\vec{i} + y\vec{j} = \vec{r}$

Problems

17. **(a)** The gradient is perpendicular to the level curves. See Figures 17.6 and 17.5. A function always increases in the direction of its gradient; this is why the values on the level curves of f and g increase as we approach the origin.

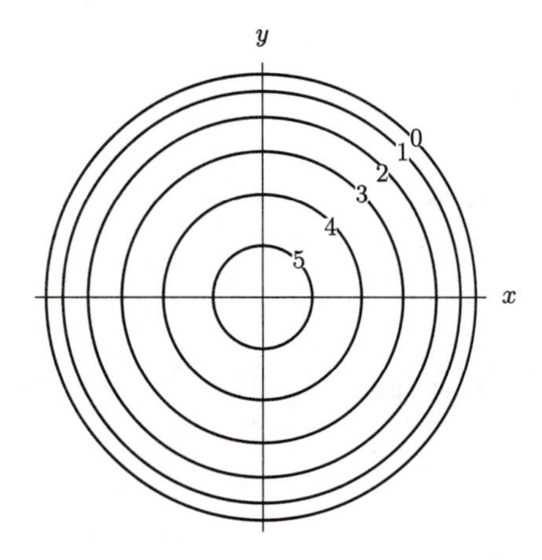

Figure 17.5: Level curves $z = f(x, y)$

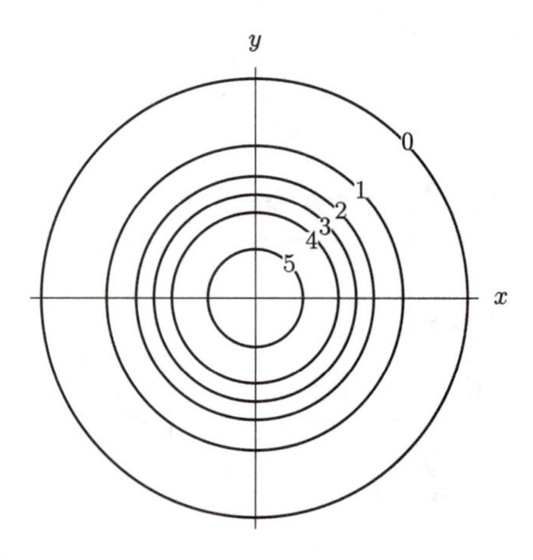

Figure 17.6: Level curves $z = g(x, y)$

(b) f climbs faster at outside, slower at center; g climbs slower at outside, faster at center:

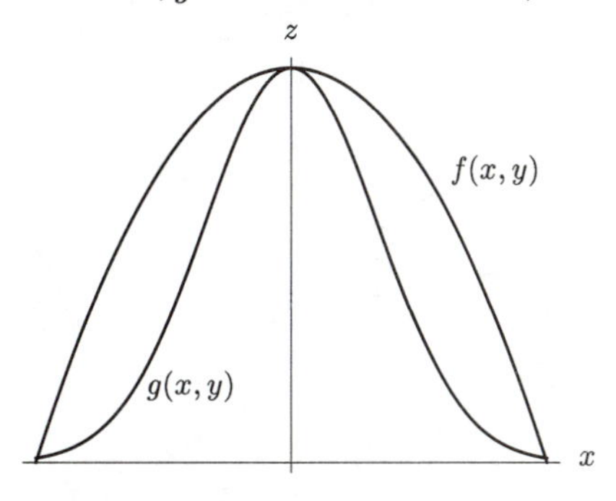

Figure 17.7

This can be understood if we notice that the magnitude of the gradient of f decreases as one approaches the origin whereas the magnitude of the gradient of g increases (at least for a while - what happens very close to the origin depends on the behavior of grad g in the region. One possibility for g is shown in Figure 17.7; the graph of g could also have a sharp peak at 0 or even blow up.)

21. (a) Since the velocity of the water is the sum of the velocities of the individual fields, then the total field should be

$$\vec{v} = \vec{v}_{\text{stream}} + \vec{v}_{\text{fountain}}.$$

It is reasonable to represent $\vec{v}_{\text{stream}}$ by the vector field $\vec{v}_{\text{stream}} = A\vec{i}$, since $A\vec{i}$ is a constant vector field flowing in the i-direction (provided $A > 0$). It is reasonable to represent $\vec{v}_{\text{fountain}}$ by

$$\vec{v}_{\text{fountain}} = K\vec{r}_r/r^2 = K(x^2+y^2)^{-1}(x\vec{i}+y\vec{j}),$$

since this is a vector field flowing radially outward (provided $K > 0$), with decreasing velocity as r gets larger. We would expect the velocity to decrease as the water from the fountain spreads out. Adding the two vector fields together, we get

$$\vec{v} = A\vec{i} + K(x^2+y^2)^{-1}(x\vec{i}+y\vec{j}), \quad A > 0, K > 0.$$

(b) The constants A and K signify the strength of the individual components of the field. A is the strength of the flow of the stream alone (in fact it is the speed of the stream), and K is the strength of the fountain acting alone.

(c)

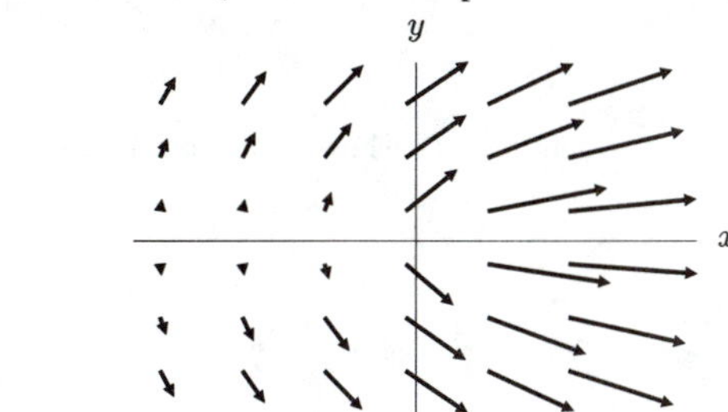

Figure 17.8: $A = 1, K = 1$

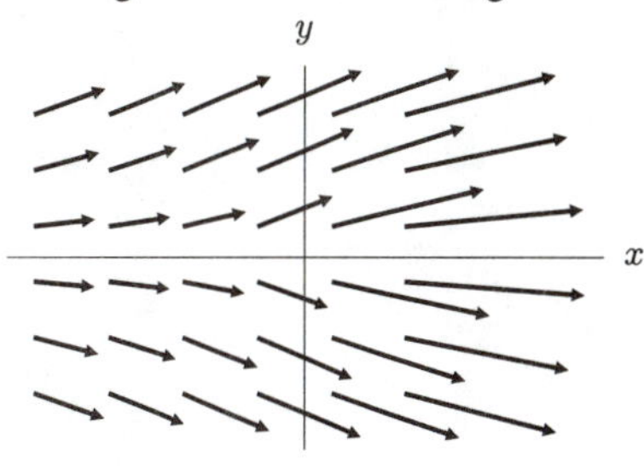

Figure 17.9: $A = 2, K = 1$

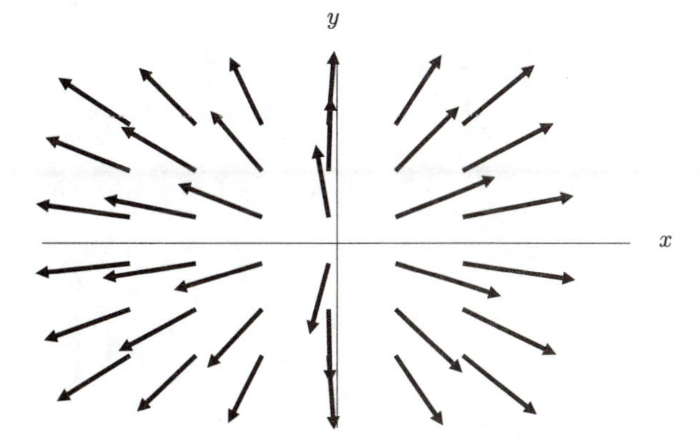

Figure 17.10: $A = 0.2, K = 2$

25. **(a)** The vector field $\vec{L} = 0\vec{F} + \vec{G} = -y\vec{i} + x\vec{j}$ is shown in Figure 17.11.

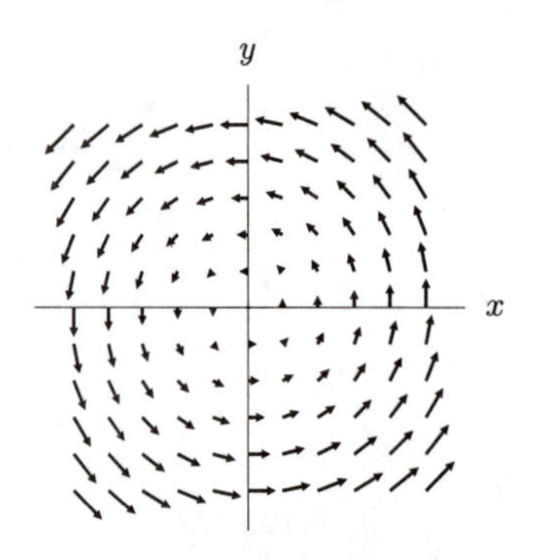

Figure 17.11

(b) The vector field $\vec{L} = a\vec{F} + \vec{G} = (ax - y)\vec{i} + (ay + x)\vec{j}$ where $a > 0$ is shown in Figure 17.12.

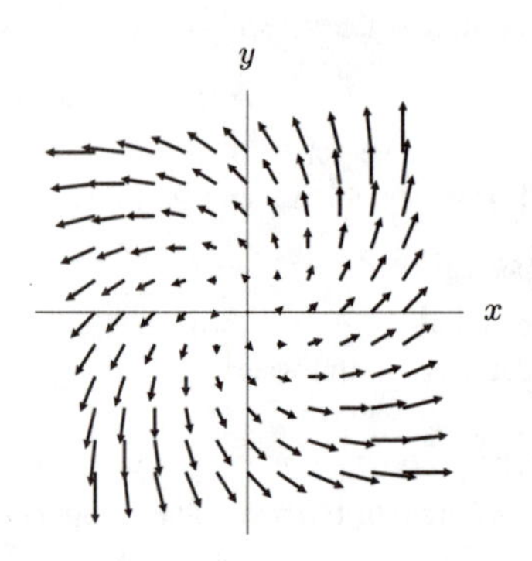

Figure 17.12

(c) The vector field $\vec{L} = a\vec{F} + \vec{G} = (ax - y)\vec{i} + (ay + x)\vec{j}$ where $a < 0$ is shown in Figure 17.13.

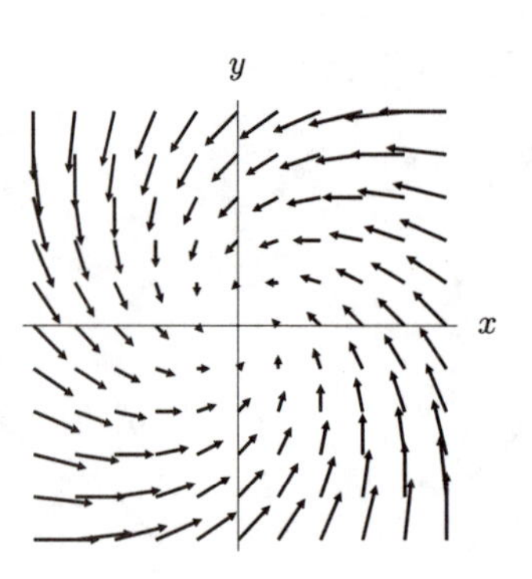

Figure 17.13

Solutions for Section 17.4

Exercises

1. Since $x'(t) = 3$ and $y'(t) = 0$, we have $x = 3t + x_0$ and $y = y_0$. Thus, the solution curves are $y =$ constant.

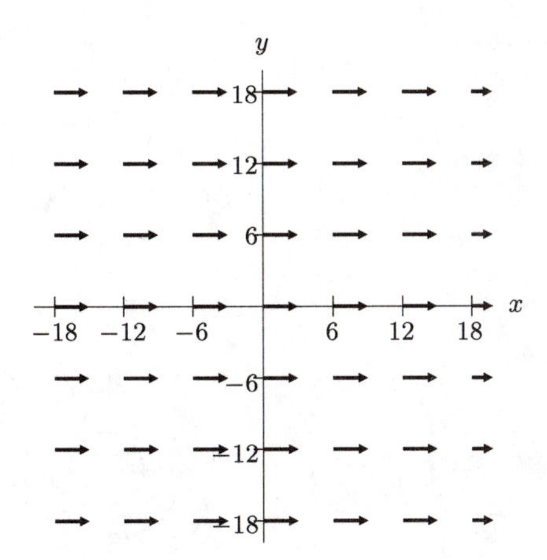

Figure 17.14: The field $\vec{v} = 3\vec{i}$

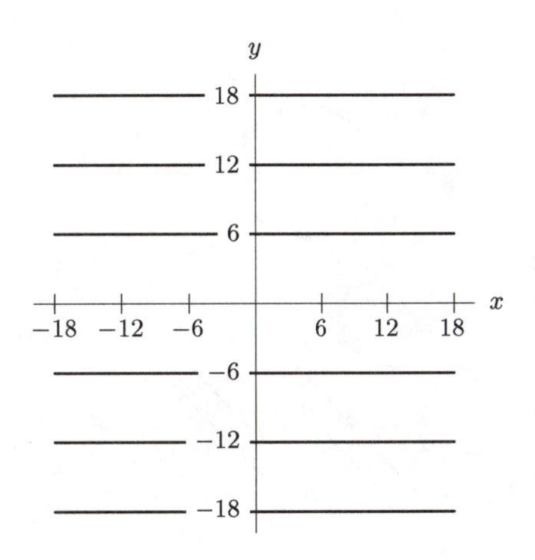

Figure 17.15: The flow $y =$constant

5.

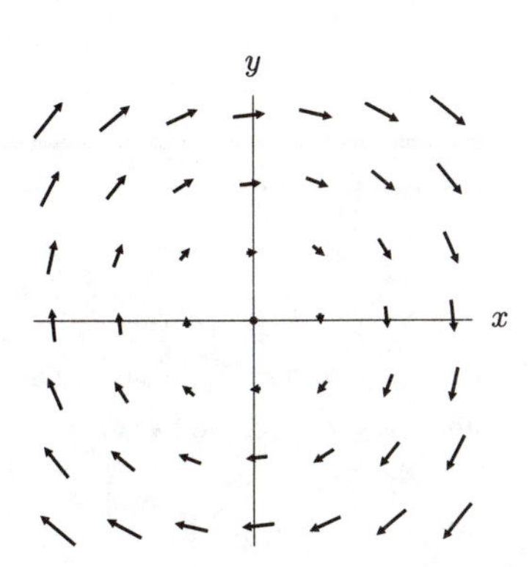

Figure 17.16: $\vec{v}(t) = y\vec{i} - x\vec{j}$

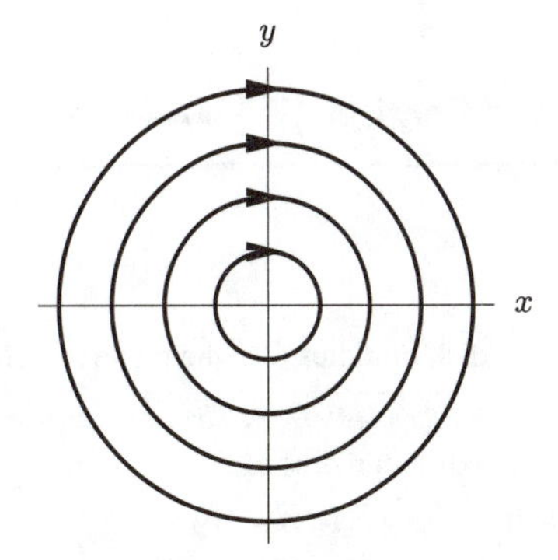

Figure 17.17: The flow $x = a\sin t$, $y = a\cos t$

As

$$\vec{v}(t) = \frac{dx}{dt}\vec{i} + \frac{dy}{dt}\vec{j},$$

the system of differential equations is

$$\begin{cases} \frac{dx}{dt} = y \\ \frac{dy}{dt} = -x. \end{cases}$$

Since

$$\frac{dx(t)}{dt} = \frac{d}{dt}[a\sin t] = a\cos t = y(t)$$

and

$$\frac{dy(t)}{dt} = \frac{d}{dt}[a\cos t] = -a\sin t = -x(t),$$

the given flow satisfies the system. By eliminating the parameter t in $x(t)$ and $y(t)$, the solution curves obtained are $x^2 + y^2 = a^2$.

9. The directions of the flow lines are as shown.

(a) III
(b) I
(c) II
(d) V
(e) VI
(f) IV

(I)
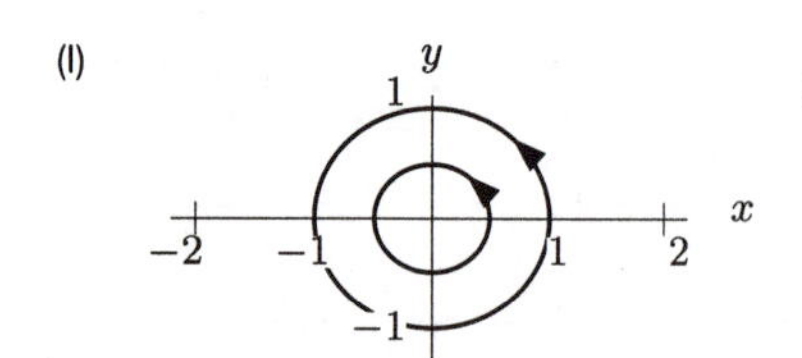

(II)
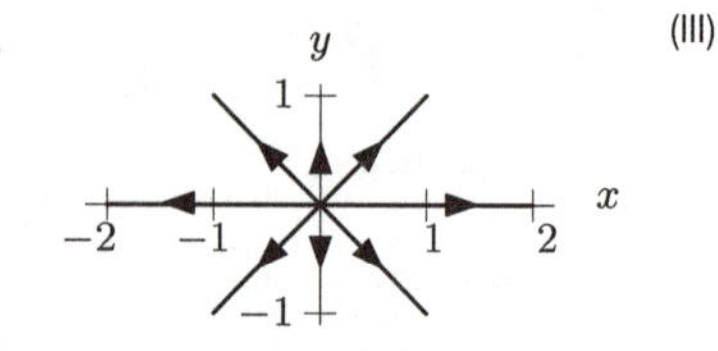

(III)
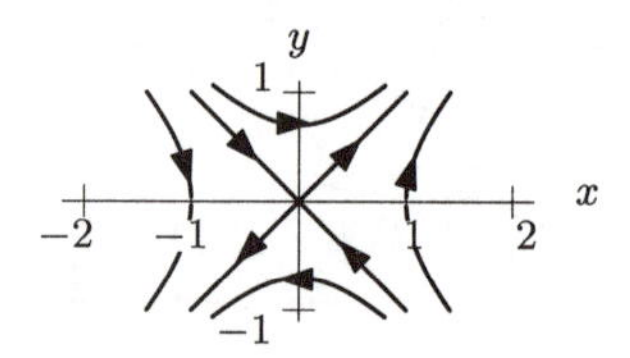

(IV)
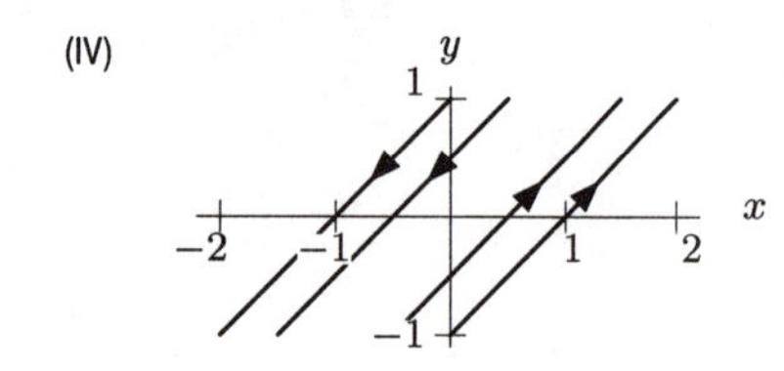

(V)
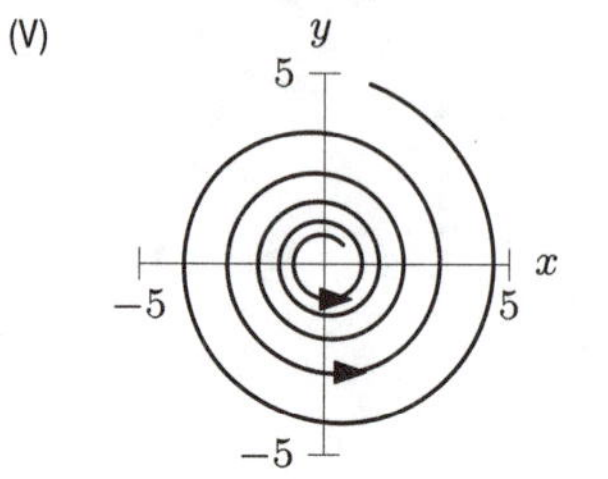

(VI)
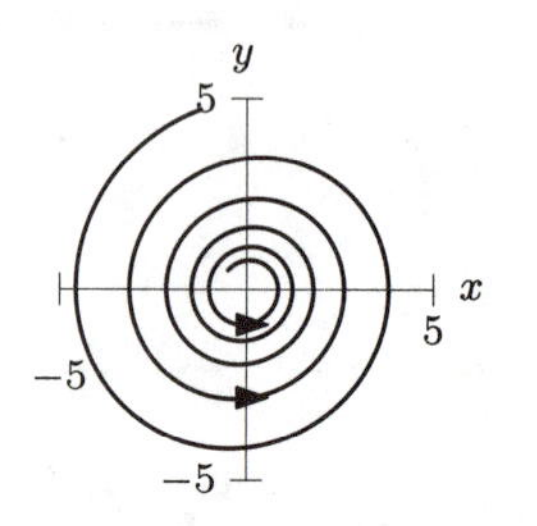

Solutions for Section 17.5

Exercises

1. A horizontal disk of radius 5 in the plane $z = 7$.

5. Since $z = r = \sqrt{x^2 + y^2}$, we have a cone around the z-axis. Since $0 \le r \le 5$, we have $0 \le z \le 5$, so the cone has height and maximum radius of 5.

9. The top half of the sphere $(z \ge 0)$.

Problems

13. The cross sections of the cylinder perpendicular to the z-axis are circles which are vertical translates of the circle $x^2+y^2 = a^2$, which is given parametrically by $x = a\cos\theta$, $y = a\sin\theta$. The vector $a\cos\theta\vec{i} + a\sin\theta\vec{j}$ traces out the circle, at any height. We get to a point on the surface by adding that vector to the vector $z\vec{k}$. Hence, the parameters are θ, with $0 \le \theta \le 2\pi$, and z, with $0 \le z \le h$. The parametric equations for the cylinder are

$$x\vec{i} + y\vec{j} + z\vec{k} = a\cos\theta\vec{i} + a\sin\theta\vec{j} + z\vec{k},$$

which can be written as

$$x = a\cos\theta, \quad y = a\sin\theta, \quad z = z.$$

17. **(a)** We want to find s and t so that

$$\begin{aligned} 2 + s &= 4 \\ 3 + s + t &= 8 \\ 4t &= 12 \end{aligned}$$

Since $s = 2$ and $t = 3$ satisfy these equations, the point $(4, 8, 12)$ lies on this plane.

(b) Are there values of s and t corresponding to the point $(1, 2, 3)$? If so, then

$$\begin{aligned} 1 &= 2 + s \\ 2 &= 3 + s + t \\ 3 &= 4t \end{aligned}$$

From the first equation we must have $s = -1$ and from the third we must have $t = 3/4$. But these values of s and t do not satisfy the second equation. Therefore, no value of s and t corresponds to the point $(1, 2, 3)$, and so $(1, 2, 3)$ is not on the plane.

21. A vertical half-circle, going from the north to south poles.

25. The parameterization for a sphere of radius a using spherical coordinates is

$$x = a \sin\phi \cos\theta, \quad y = a \sin\phi \sin\theta, \quad z = a\cos\phi.$$

Think of the ellipsoid as a sphere whose radius is different along each axis and you get the parameterization:

$$\begin{cases} x = a\sin\phi\cos\theta, & 0 \le \phi \le \pi, \\ y = b\sin\phi\sin\theta, & 0 \le \theta \le 2\pi, \\ z = c\cos\phi. \end{cases}$$

To check this parameterization, substitute into the equation for the ellipsoid:

$$\begin{aligned} \frac{x^2}{a^2} + \frac{y^2}{b^2} + \frac{z^2}{c^2} &= \frac{a^2\sin^2\phi\cos^2\theta}{a^2} + \frac{b^2\sin^2\phi\sin^2\theta}{b^2} + \frac{c^2\cos^2\phi}{c^2} \\ &= \sin^2\phi(\cos^2\theta + \sin^2\theta) + \cos^2\phi = 1. \end{aligned}$$

29. **(a)** The cone of height h, maximum radius a, vertex at the origin and opening upward is shown in Figure 17.18.

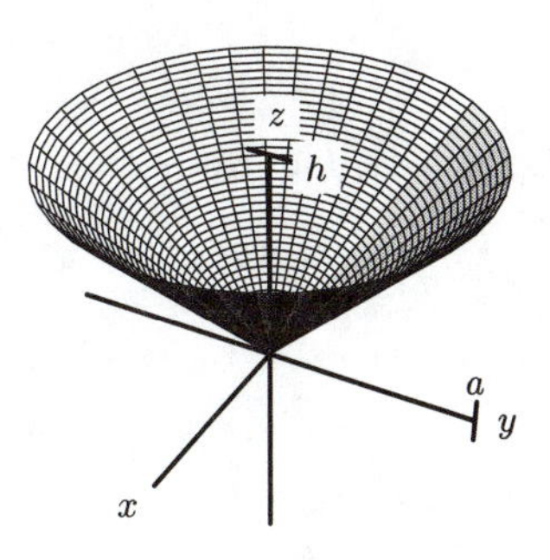

Figure 17.18

By similar triangles, we have

$$\frac{r}{z} = \frac{a}{h},$$

so

$$z = \frac{hr}{a}.$$

Therefore, one parameterization is

$$\begin{aligned} x &= r\cos\theta, & 0 \le r \le a, \\ y &= r\sin\theta, & 0 \le \theta < 2\pi, \\ z &= \frac{hr}{a}. \end{aligned}$$

(b) Since $r = az/h$, we can write the parameterization in part (a) as

$$\begin{aligned} x &= \frac{az}{h}\cos\theta, & 0 \le z \le h, \\ y &= \frac{az}{h}\sin\theta, & 0 \le \theta < 2\pi, \\ z &= z. \end{aligned}$$

33. **(a)** From the first two equations we get:

$$s = \frac{x+y}{2}, \qquad t = \frac{x-y}{2}.$$

Hence the equation of our surface is:

$$z = \left(\frac{x+y}{2}\right)^2 + \left(\frac{x-y}{2}\right)^2 = \frac{x^2}{2} + \frac{y^2}{2},$$

which is the equation of a paraboloid.

The conditions: $0 \le s \le 1$, $0 \le t \le 1$ are equivalent to: $0 \le x+y \le 2$, $0 \le x-y \le 2$. So our surface is defined by:

$$z = \frac{x^2}{2} + \frac{y^2}{2}, \qquad 0 \le x+y \le 2 \quad 0 \le x-y \le 2$$

(b) The surface is shown in Figure 17.19.

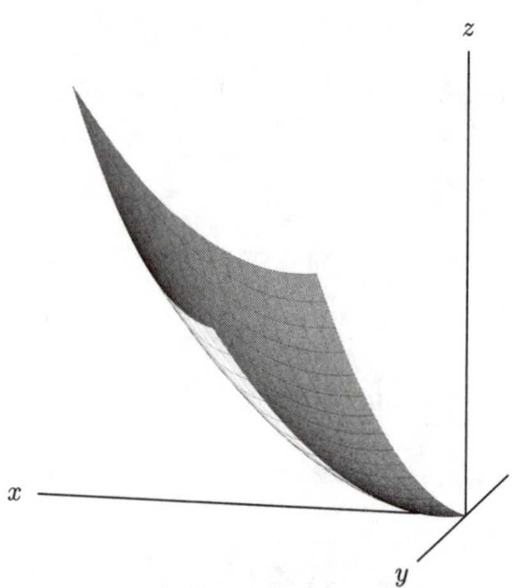

Figure 17.19: The surface $x = s + t$, $y = s - t$, $z = s^2 + t^2$ for $0 \le s \le 1$, $0 \le t \le 1$

37. The plane in which the circle lies is parameterized by

$$\vec{r}(p,q) = x_0\vec{i} + y_0\vec{j} + z_0\vec{k} + p\vec{u} + q\vec{v}.$$

Because $\vec{u}$ and $\vec{v}$ are perpendicular unit vectors, the parameters p and q establish a rectangular coordinate system on this plane exactly analogous to the usual xy-coordinate system, with $(p,q) = (0,0)$ corresponding to the point (x_0, y_0, z_0). Thus the circle we want to describe, which is the circle of radius a centered at $(p,q) = (0,0)$, can be parameterized by

$$p = a\cos t, \qquad q = a\sin t.$$

Substituting into the equation of the plane gives the desired parameterization of the circle in 3-space,

$$\vec{r}(t) = x_0\vec{i} + y_0\vec{j} + z_0\vec{k} + a\cos t\vec{u} + a\sin t\vec{v},$$

where $0 \le t \le 2\pi$.

Solutions for Chapter 17 Review

Exercises

1. $x = t, y = 5$.

5. The vector $(\vec{i} + 2\vec{j} + 5\vec{k}) - (2\vec{i} - \vec{j} + 4\vec{k}) = -\vec{i} + 3\vec{j} + \vec{k}$ is parallel to the line, so a possible parameterization is

$$x = 2 - t, \quad y = -1 + 3t, \quad z = 4 + t.$$

9. Since the circle has radius 3, the equation must be of the form $x = 3\cos t, y = 5, z = 3\sin t$. But since the circle is being viewed from farther out on the y-axis, the circle we have now would be seen going clockwise. To correct this, we add a negative to the third component, giving us the equation $x = 3\cos t, y = 5, z = -3\sin t$.

13. The vector field points in a clockwise direction around the origin. Since

$$\left\| \left(\frac{y}{\sqrt{x^2+y^2}}\right)\vec{i} - \left(\frac{x}{\sqrt{x^2+y^2}}\right)\vec{j} \right\| = \frac{\sqrt{x^2+y^2}}{\sqrt{x^2+y^2}} = 1$$

the length of the vectors is constant everywhere.

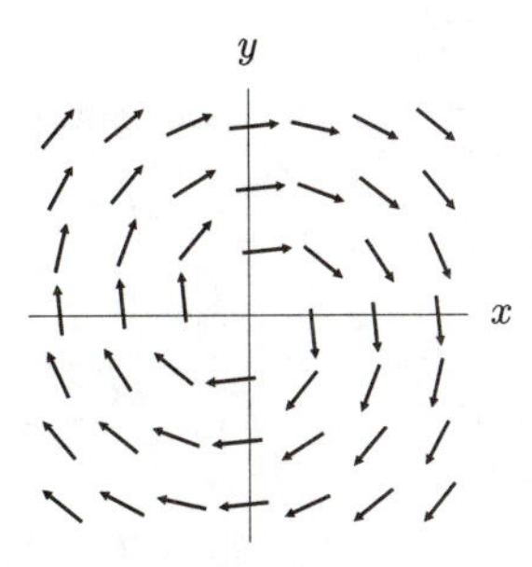

Figure 17.20

Problems

17. (a) (I) has radius 1 and traces out a complete circle, so $I = C_4$.
(II) has radius 2 and traces out the top half of a circle, so $II = C_1$
(III) has radius 1 and traces out a quarter circle, so $III = C_2$.
(IV) has radius 2 and traces out the bottom half of a circle, so $IV = C_6$.

(b) C_3 has radius $1/2$ and traces out a half circle below the x-axis, so

$$\vec{r} = 0.5\cos t\vec{i} - 0.5\sin t\vec{j}\,.$$

C_5 has radius 2 and traces out a quarter circle below the x-axis starting at the point $(-2, 0)$. Thus we have

$$\vec{r} = -2\cos(t/2)\vec{i} - 2\sin(t/2)\vec{j}\,.$$

21. (a) Separate the ant's path into three parts: from $(0, 0)$ to $(1, 0)$ along the x-axis; from $(1, 0)$ to $(0, 1)$ via the circle; and from $(0, 1)$ to $(0, 0)$ along the y-axis. (See Figure 17.21.) The lengths of the paths are 1, $\frac{2\pi}{4} = \frac{\pi}{2}$, and 1 respectively. Thus, the time it takes for the ant to travel the three paths are (using the formula $t = \frac{d}{v}$) $\frac{1}{2}$, $\frac{1}{3}$, and $\frac{1}{2}$ seconds.

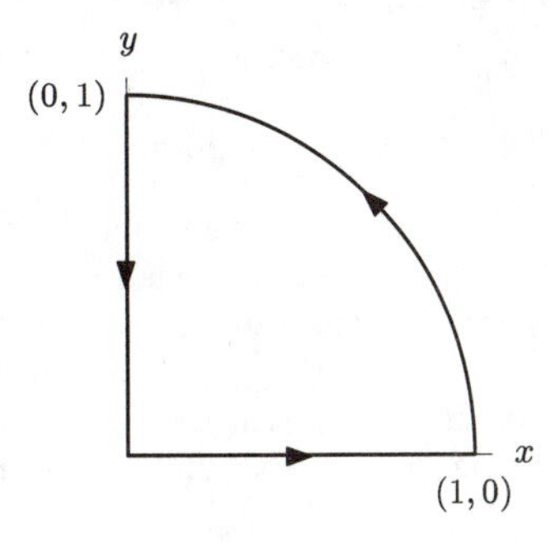

Figure 17.21

From $t = 0$ to $t = \frac{1}{2}$, the ant is heading toward $(1, 0)$ so its coordinate is $(2t, 0)$. From $t = \frac{1}{2}$ to $t = \frac{1}{2} + \frac{1}{3} = \frac{5}{6}$, the ant is veering to the left and heading toward $(0, 1)$. At $t = \frac{1}{2}$, it is at $(1, 0)$ and at $t = \frac{5}{6}$, it is at $(0, 1)$. Thus its position is $(\cos[\frac{3\pi}{2}(t - \frac{1}{2})], \sin[\frac{3\pi}{2}(t - \frac{1}{2})])$. Finally, from $t = \frac{5}{6}$ to $t = \frac{5}{6} + \frac{1}{2} = \frac{4}{3}$, the ant is headed home. Its coordinates are $(0, -2(t - \frac{4}{3}))$.

In summary, the function expressing the ant's coordinates is

$$(x(t), y(t)) = \begin{cases} (2t, 0) & \text{when } 0 \le t \le \frac{1}{2} \\ \left(\cos(\frac{3\pi}{2}(t - \frac{1}{2})), \sin(\frac{3\pi}{2}(t - \frac{1}{2}))\right) & \text{when } \frac{1}{2} < t \le \frac{5}{6} \\ (0, -2(t - \frac{4}{3})) & \text{when } \frac{5}{6} \le t \le \frac{4}{3}. \end{cases}$$

(b) To do the reverse path, observe that we can reverse the ant's path by interchanging the x and y coordinates (flipping it with respect to the line $y = x$), so the function is

$$(x(t), y(t)) = \begin{cases} (0, 2t) & \text{when } 0 \le t \le \frac{1}{2} \\ \left(\sin(\frac{3\pi}{2}(t - \frac{1}{2})), \cos(\frac{3\pi}{2}(t - \frac{1}{2}))\right) & \text{when } \frac{1}{2} < t \le \frac{5}{6} \\ (-2(t - \frac{4}{3}), 0) & \text{when } \frac{5}{6} < t \le \frac{4}{3}. \end{cases}$$

25. (a) The equation of the sphere is $x^2 + y^2 + z^2 = 1$. Substituting the parameterization gives

$$(x(t))^2+(y(t))^2+(z(t))^2 = \cos^4 t+\sin^2 t\cos^2 t+\sin^2 t = (\cos^2 t)(\cos^2 t+\sin^2 t)+\sin^2 t = \cos^2 t+\sin^2 t = 1.$$

Therefore the curve lies on this sphere.

(b) If $t = \pi/4$, we have $x(\pi/4) = 1/2$, $y(\pi/4) = 1/2$, $z(\pi/4) = 1/\sqrt{2}$. The gradient vector to the sphere at this point is perpendicular to the sphere and the curve. Since $\text{grad}\, f = 2x\vec{i} + 2y\vec{j} + 2z\vec{k}$, we have

$$\text{Normal} = \text{grad}\, f\left(\frac{1}{2}, \frac{1}{2}, \frac{1}{\sqrt{2}}\right) = 2\left(\frac{1}{2}\right)\vec{i} + 2\left(\frac{1}{2}\right)\vec{j} + 2\left(\frac{1}{\sqrt{2}}\right)\vec{k} = \vec{i} + \vec{j} + \sqrt{2}\vec{k}.$$

(c) A tangent vector is $x'(t)\vec{i} + y'(t)\vec{j} + z'(t)\vec{k} = (-2\cos t\sin t)\vec{i} + (\cos^2 t - \sin^2 t)\vec{j} + \cos t\vec{k}$. At $t = \pi/4$, we have

$$\text{Tangent vector} = -\vec{i} + \frac{1}{\sqrt{2}}\vec{k}.$$

29. Let's place the coordinate system so that the origin is at the center of the circle of radius R on which the object is moving. If $\theta(t)$ is the polar coordinate angle of the object at time t, the position vector of the object is given by

$$\vec{r}(t) = R\cos(\theta(t))\vec{i} + R\sin(\theta(t))\vec{j}.$$

(a) The velocity vector is given by the derivative

$$\vec{r}' = -R\theta'\sin\theta\vec{i} + R\theta'\cos\theta\vec{j} = R\theta'(-\sin\theta\vec{i} + \cos\theta\vec{j})$$

The velocity $\vec{r}'$ is perpendicular to the radius vector $\vec{r}$ from the center of the circle to the object because

$$\begin{aligned} \vec{r}' \cdot \vec{r} &= R\theta'\left(-\sin\theta\vec{i} + \cos\theta\vec{j}\right) \cdot \left(R\cos\theta\vec{i} + R\sin\theta\vec{j}\right) \\ &= R^2\theta'(-\sin\theta\cos\theta + \cos\theta\sin\theta) = 0. \end{aligned}$$

So the velocity vector, $\vec{r}'$, is tangent to the circle.

(b) The acceleration vector is given by the second derivative

$$\vec{r}'' = R\theta''(-\sin\theta\vec{i} + \cos\theta\vec{j}) - R(\theta')^2(\cos\theta\vec{i} + \sin\theta\vec{j})$$

The acceleration vector $\vec{r}''$ is expressed as the sum of two vector components. Since $\vec{r} = R\cos\theta\vec{i} + R\sin\theta\vec{j}$, the component $-R(\theta')^2(\cos\theta\vec{i} + \sin\theta\vec{j})$ is in the direction opposite to $\vec{r}$, that is, toward the center of the circle. The component $R\theta''(-\sin\theta\vec{i} + \cos\theta\vec{j})$ is perpendicular to $\vec{r}$ since the dot product is $(-\sin\theta\vec{i} + \cos\theta\vec{j}) \cdot (\cos\theta\vec{i} + \sin\theta\vec{j}) = 0$. This means that $R\theta''\left(-\sin\theta\vec{i} + \cos\theta\vec{j}\right)$ is in the direction tangent to the circle. The sum $\vec{r}''$ is not in general directed toward the center of the circle. The acceleration is only directed toward the center of the circle in the case where $\theta''(t) = 0$, as it is for example in the case $\theta(t) = \omega t$ of uniform circular motion.

33. This corresponds to area C in Figure 17.22.

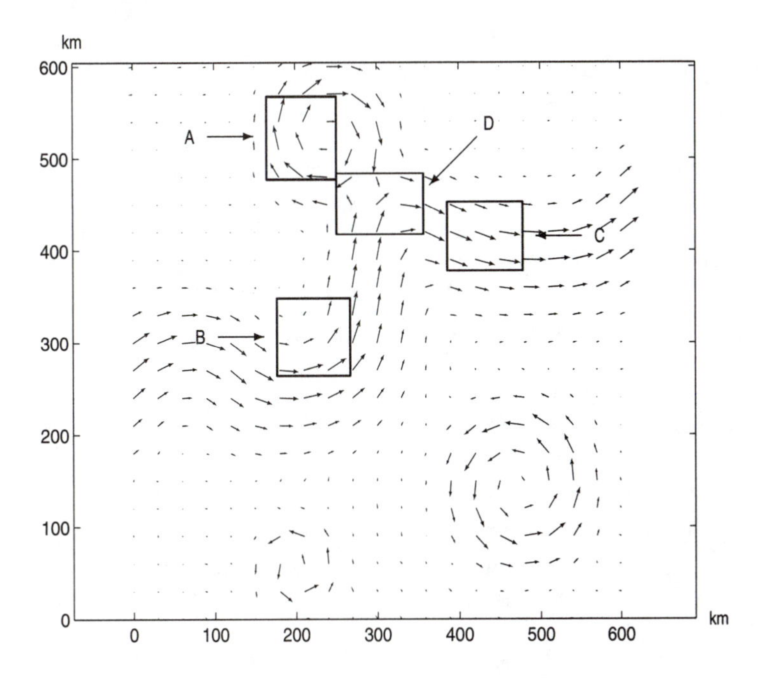

Figure 17.22

37. The sphere $x^2 + y^2 + z^2 = 1$ is shown in Figure 17.23.

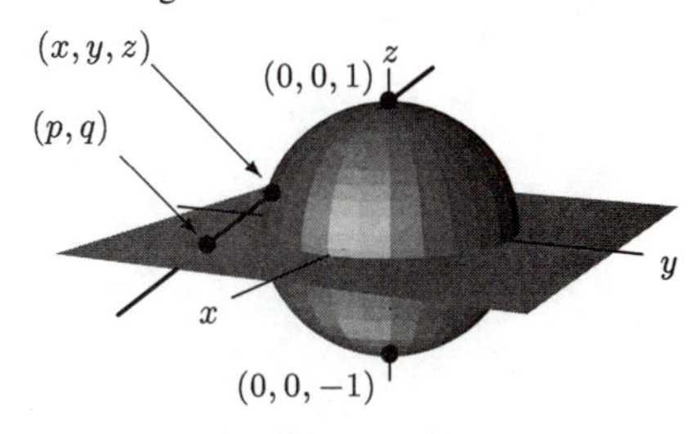

Figure 17.23

(a) The origin corresponds to the south pole.
(b) The circle $x^2 + y^2 = 1$ corresponds to the equator.
(c) We get all the points of the sphere by this parameterization except the north pole itself.
(d) $x^2 + y^2 > 1$ corresponds to the upper hemisphere.
(e) $x^2 + y^2 < 1$ corresponds to the lower hemisphere.

CAS Challenge Problems

41. Answers may differ depending on the method and CAS used.

(a) Using a CAS to solve for x and y in terms of z and letting $z = t$, we get $x = \frac{20}{13} - \frac{6t}{13}, y = \frac{-1}{13} - \frac{t}{13}, z = t$.
(b) Using a CAS to solve for y and z in terms of x and letting $x = t$, we get $x = t, y = \frac{1}{6}(-2 - 2t + 3t^2), z = \frac{1}{6}(20 - 10t - 3t^2)$.
(c) Using a CAS to solve for x and z in terms of y, we get two solutions

$$x = \sqrt{2 - t^2}, \quad y = t, \quad z = 5 + 5t - 3\sqrt{2 - t^2}$$

and

$$x = -\sqrt{2 - t^2}, \quad y = t, \quad z = 5 + 5t + 3\sqrt{2 - t^2}$$

Each of these is a parameterization of one half of the intersection curve.

CHECK YOUR UNDERSTANDING

1. False. The y coordinate is zero when $t = 0$, but when $t = 0$ we have $x = 2$ so the curve never passes through $(0, 0)$.

5. False. When $t = 0$, we have $(x, y) = (0, -1)$. When $t = \pi/2$, we have $(x, y) = (-1, 0)$. Thus the circle is being traced out clockwise.

9. True. To find an intersection point, we look for values of s and t that make the coordinates in the first line the same as the coordinates in the second. Setting $x = t$ and $x = 2s$ equal, we see that $t = 2s$. Setting $y = 2 + t$ equal to $y = 1 - s$, we see that $t = -1 - s$. Solving both $t = 2s$ and $t = -1 - s$ yields $t = -\frac{2}{3}, s = -\frac{1}{3}$. These values of s and t will give equal x and y coordinates on both lines. We need to check if the z coordinates are equal also. In the first line, setting $t = -\frac{2}{3}$ gives $z = \frac{7}{3}$. In the second line, setting $s = -\frac{1}{3}$ gives $z = -\frac{1}{3}$. As these are not the same, the lines do not intersect.

13. False. The velocity vector is $\vec{v}(t) = \vec{r}\,'(t) = 2t\vec{i} - \vec{j}$. Then $\vec{v}(-1) = -2\vec{i} - \vec{j}$ and $\vec{v}(1) = 2\vec{i} - \vec{j}$, which are not equal.

17. False. As a counterexample, consider the curve $\vec{r}(t) = t^2\vec{i} + t^2\vec{j}$ for $0 \leq t \leq 1$. In this case, when t is replaced by $-t$, the parameterization is the same, and is not reversed.

21. True, since the vectors $x\vec{j}$ are parallel to the y-axis.

25. True. Any flow line which stays in the first quadrant has $x, y \to \infty$.

29. True. If (x, y) were a point where the y-coordinate along a flow line reached a relative maximum, then the tangent vector to the flow line, namely $\vec{F}(x, y)$, there would have to be horizontal (or $\vec{0}$), that is its $\vec{j}$ component would have to be 0. But the $\vec{j}$ component of $\vec{F}$ is always 2.

33. False. There is only one parameter, s. The equations parameterize a line.

37. True. If the surface is parameterized by $\vec{r}(s, t)$ and the point has parameters (s_0, t_0) then the parameter curves $\vec{r}(s_0, t)$ and $\vec{r}(s, t_0)$.

41. False. Suppose $\vec{r}(t) = t\vec{i} + t\vec{j}$. Then $\vec{r}\,'(t) = \vec{i} + \vec{j}$ and

$$\vec{r}\,'(t) \cdot \vec{r}(t) = (t\vec{i} + t\vec{j}) \cdot (\vec{i} + \vec{j}) = 2t.$$

So $\vec{r}\,'(t) \cdot \vec{r}(t) \neq 0$ for $t \neq 0$.

CHAPTER EIGHTEEN

Solutions for Section 18.1

Exercises

1. Positive, because the vectors are longer on the portion of the path that goes in the same direction as the vector field.

5. Since $\vec{F}$ is perpendicular to the curve at every point along it,

$$\int_C \vec{F} \cdot d\vec{r} = 0.$$

9. At every point on the path, $\vec{F}$ is parallel to $\Delta\vec{r}$. Suppose r is the distance from the point (x, y) to the origin, so $\|\vec{r}\| = r$. Then $\vec{F} \cdot \Delta\vec{r} = \|\vec{F}\|\|\Delta\vec{r}\| = r\Delta r$. At the start of the path, $r = \sqrt{2^2 + 2^2} = 2\sqrt{2}$ and at the end $r = 6\sqrt{2}$. Thus,

$$\int_C \vec{F} \cdot d\vec{r} = \int_{2\sqrt{2}}^{6\sqrt{2}} r dr = \left.\frac{r^2}{2}\right|_{2\sqrt{2}}^{6\sqrt{2}} = 32.$$

Problems

13. Since it appears that C_1 is everywhere perpendicular to the vector field, all of the dot products in the line integral are zero, hence $\int_{C_1} \vec{F} \cdot d\vec{r} \approx 0$. Along the path C_2 the dot products of $\vec{F}$ with $\Delta\vec{r_i}$ are all positive, so their sum is positive and we have $\int_{C_1} \vec{F} \cdot d\vec{r} < \int_{C_2} \vec{F} \cdot d\vec{r}$. For C_3 the vectors $\Delta\vec{r_i}$ are in the opposite direction to the vectors of $\vec{F}$, so the dot products $\vec{F} \cdot \Delta\vec{r_i}$ are all negative; so, $\int_{C_3} \vec{F} \cdot d\vec{r} < 0$. Thus, we have

$$\int_{C_3} \vec{F} \cdot d\vec{r} < \int_{C_1} \vec{F} \cdot d\vec{r} < \int_{C_2} \vec{F} \cdot d\vec{r}$$

17. Since it does not depend on y, this vector field is constant along vertical lines, $x =$ constant. Now let us consider two points P and Q on C_1 which lie on the same vertical line. Because C_1 is symmetric with respect to the x-axis, the tangent vectors at P and Q will be symmetric with respect to the vertical axis so their sum is a vertical vector. But $\vec{F}$ has only horizontal component and thus $\vec{F} \cdot (\Delta\vec{r}(P) + \Delta\vec{r}(Q)) = 0$. As $\vec{F}$ is constant along vertical lines (so $\vec{F}(P) = \vec{F}(Q)$), we obtain

$$\vec{F}(P) \cdot \Delta\vec{r}(P) + \vec{F}(Q) \cdot \Delta\vec{r}(Q) = 0.$$

Summing these products and making $\|\Delta\vec{r}\| \to 0$ gives us

$$\int_{C_1} \vec{F} \cdot d\vec{r} = 0.$$

The same thing happens on C_3, so $\int_{C_3} \vec{F} \cdot d\vec{r} = 0$.

Now let P be on C_2 and Q on C_4 lying on the same vertical line. The respective tangent vectors are symmetric with respect to the vertical axis hence they add up to a vertical vector and a similar argument as before gives

$$\vec{F}(P) \cdot \Delta\vec{r}(P) + \vec{F}(Q) \cdot \Delta\vec{r}(Q) = 0$$

and

$$\int_{C_2} \vec{F} \cdot d\vec{r} + \int_{C_4} \vec{F} \cdot d\vec{r} = 0$$

and so

$$\int_C \vec{F} \cdot d\vec{r} = 0.$$

See Figure 18.1.

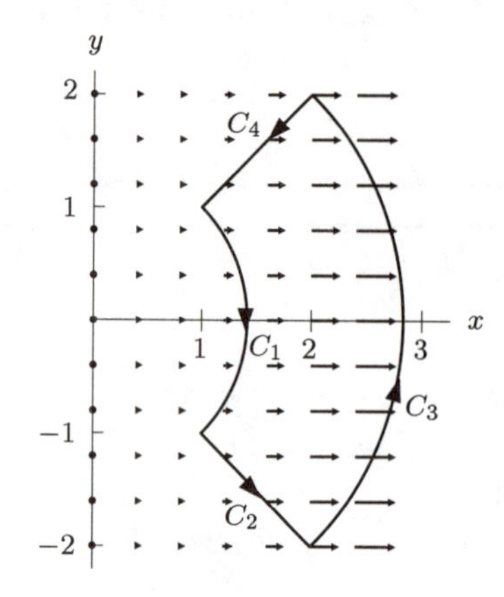

Figure 18.1

21. Suppose $\int_C \vec{F} \cdot d\vec{r} = 0$ for every closed curve C. Pick any two fixed points P_1, P_2 and curves C_1, C_2 each going from P_1 to P_2. See Figure 18.2. Define $-C_2$ to be the same curve as C_2 except in the opposite direction. Therefore, the curve formed by traversing C_1, followed by C_2 in the opposite direction, written as $C_1 - C_2$, is a closed curve, so by our assumption, $\int_{C_1-C_2} \vec{F} \cdot d\vec{r} = 0$. However, we can write

$$\int_{C_1-C_2} \vec{F} \cdot d\vec{r} = \int_{C_1} \vec{F} \cdot d\vec{r} - \int_{C_2} \vec{F} \cdot d\vec{r}$$

since C_2 and $-C_2$ are the same except for direction. Therefore,

$$\int_{C_1} \vec{F} \cdot d\vec{r} - \int_{C_2} \vec{F} \cdot d\vec{r} = 0,$$

so

$$\int_{C_1} \vec{F} \cdot d\vec{r} = \int_{C_2} \vec{F} \cdot d\vec{r}.$$

Since C_1 and C_2 are any two curves with the endpoints P_1, P_2, this gives the desired result – namely, that fixing endpoints and direction uniquely determines the value of $\int_C \vec{F} \cdot d\vec{r}$. In other words, the value of the integral $\int_C \vec{F} \cdot d\vec{r}$ does not depend on the path taken. We say the line integral is *path-independent.*

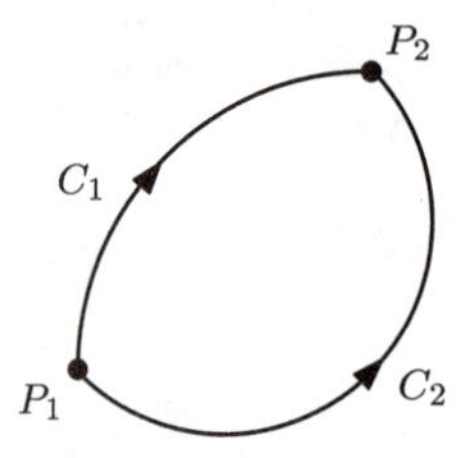

Figure 18.2

25. Let $r = \|\vec{r}\|$. Since $\Delta\vec{r}$ points outward, in the opposite direction to $\vec{F}$, we expect a negative answer. We take the upper limit to be $r = \infty$, so the integral is improper.

$$\begin{aligned}\int_C \vec{F} \cdot d\vec{r} &= \int_C -\frac{GMm\vec{r}}{r^3} \cdot d\vec{r} = \int_{8000}^{\infty} -\frac{GMm}{r^2} dr \\ &= \lim_{b\to\infty} \int_{8000}^{b} -\frac{GMm}{r^2} dr = \lim_{b\to\infty} \left.\frac{GMm}{r}\right|_{8000}^{b} = \lim_{b\to\infty} GMm\left(\frac{1}{b} - \frac{1}{8000}\right) \\ &= -\frac{GMm}{8000}\end{aligned}$$

29. (a) Suppose P is b units from the origin. Then P can be reached by a path, C, consisting of two pieces, C_1 and C_2, one lying on the sphere of radius a and one going straight along a line radiating from the origin (see Figure 18.3). We have $\vec{E} \cdot \Delta\vec{r} = 0$ on C_1, and, writing $r = \|\vec{r}\|$, we have $\vec{E} \cdot \Delta\vec{r} = \|\vec{E}\|\Delta r$ on C_2, so

$$\begin{aligned}\phi(P) &= -\int_C \vec{E} \cdot d\vec{r} = -\int_{C_1} \vec{E} \cdot d\vec{r} - \int_{C_2} \vec{E} \cdot d\vec{r} \\ &= 0 - \int_a^b \|\vec{E}\|\, dr = 0 - \int_a^b \frac{Q}{4\pi\epsilon}\frac{1}{r^2}\, dr \\ &= \frac{Q}{4\pi\epsilon}\frac{1}{r}\bigg|_a^b = \frac{Q}{4\pi\epsilon}\frac{1}{b} - \frac{Q}{4\pi\epsilon}\frac{1}{a}.\end{aligned}$$

Let P be the point with position vector $\vec{r}$. Then

$$\phi(\vec{r}) = -\frac{Q}{4\pi\epsilon}\frac{1}{a} + \frac{Q}{4\pi\epsilon}\frac{1}{\|\vec{r}\|}.$$

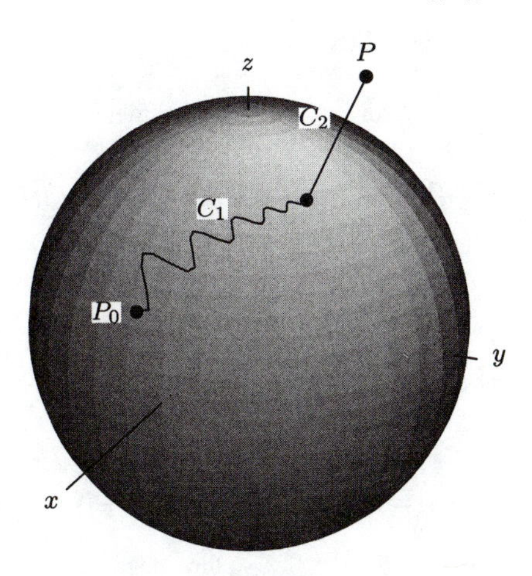

Figure 18.3

(b) If we let $a \to \infty$ in the formula for ϕ, the first term goes to zero and we get the simpler expression

$$\phi(\vec{r}) = \frac{Q}{4\pi\epsilon}\frac{1}{\|\vec{r}\|}.$$

Solutions for Section 18.2

Exercises

1. The curve C is parameterized by $(x, y) = (t, t)$ for $0 \le t \le 3$. Thus,

$$\int_C \vec{F} \cdot d\vec{r} = \int_0^3 (t\vec{i} + t\vec{j}) \cdot (\vec{i} + \vec{j})dt = \int_0^3 2tdt = t^2\bigg|_0^3 = 9.$$

5. Use $x(t) = t$, $y(t) = t^2$, so $x'(t) = 1$, $y'(t) = 2t$, with $0 \le t \le 2$. Then

$$\begin{aligned}\int \vec{F} \cdot d\vec{r} &= \int_0^2 (-t^2 \sin t\vec{i} + \cos t\vec{j}) \cdot (\vec{i} + 2t\vec{j})\, dt \\ &= \int_0^2 (-t^2 \sin t + 2t\cos t)\, dt = t^2 \cos t\Big|_0^2 = 4\cos 2.\end{aligned}$$

9. The path can be broken into three line segments: C_1, from $(1,0)$ to $(-1,0)$, and C_2, from $(-1,0)$ to $(0,1)$, and C_3, from $(0,1)$ to $(1,0)$. (See Figure 18.4.)

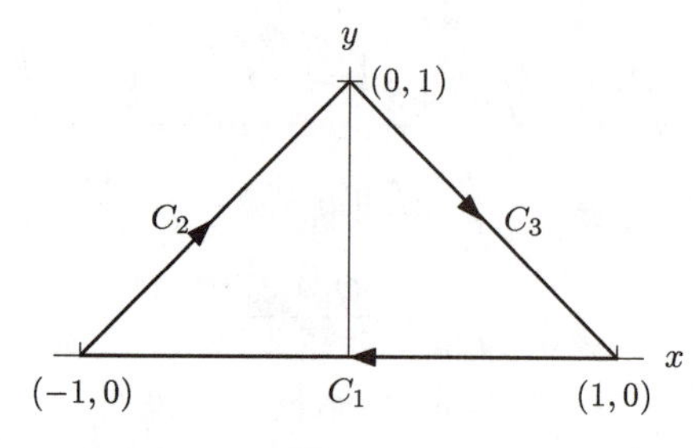

Figure 18.4

Along C_1 we have $y = 0$ so the vector field $xy\vec{i} + (x-y)\vec{j}$ is perpendicular to C_1; Thus, the line integral along C_1 is 0.

C_2 can be parameterized by $(-1+t, t)$, for $0 \le t \le 1$ so the integral is

$$\begin{aligned}
\int_{C_2} \vec{F} \cdot d\vec{r} &= \int_0^1 \vec{F}(-1+t,t) \cdot (\vec{i} + \vec{j})\, dt \\
&= \int_0^1 [t(-1+t)\vec{i} + (-1)\vec{j}] \cdot (\vec{i} + \vec{j})\, dt \\
&= \int_0^1 (-t + t^2 - 1)\, dt \\
&= (-t^2/2 + t^3/3 - t)\Big|_0^1 \\
&= -1/2 + 1/3 - 1 - (0+0+0) = -7/6
\end{aligned}$$

C_3 can be parameterized by $(t, 1-t)$, for $0 \le t \le 1$ so the integral is

$$\begin{aligned}
\int_{C_3} \vec{F} \cdot d\vec{r} &= \int_0^1 \vec{F}(t, 1-t) \cdot (\vec{i} - \vec{j})\, dt \\
&= \int_0^1 (t(1-t)\vec{i} + (2t-1)\vec{j}) \cdot (\vec{i} - \vec{j})\, dt \\
&= \int_0^1 (-t^2 - t + 1)\, dt \\
&= (-t^3/3 - t^2/2 + t)\Big|_0^1 \\
&= -1/3 - 1/2 + 1 - (0+0+0) = 1/6
\end{aligned}$$

So the total line integral is

$$\int_C \vec{F} \cdot d\vec{r} = \int_{C_1} \vec{F} \cdot d\vec{r} + \int_{C_2} \vec{F} \cdot d\vec{r} + \int_{C_3} \vec{F} \cdot d\vec{r} = 0 + (-\frac{7}{6}) + \frac{1}{6} = -1$$

Problems

13. (a) Since $\vec{r}(t) = t\vec{i} + t^2\vec{j}$, we have $\vec{r}\,'(t) = \vec{i} + 2t\vec{j}$. Thus,

$$\begin{aligned}
\int_C \vec{F} \cdot d\vec{r} &= \int_0^1 \vec{F}(t,t^2) \cdot (\vec{i} + 2t\vec{j})\, dt \\
&= \int_0^1 [(3t - t^2)\vec{i} + t\vec{j}] \cdot (\vec{i} + 2t\vec{j})\, dt
\end{aligned}$$

$$= \int_0^1 (3t + t^2)\,dt$$

$$= (\frac{3t^2}{2} + \frac{t^3}{3})\Big|_0^1 = \frac{3}{2} + \frac{1}{3} - (0+0) = \frac{11}{6}$$

(b) Since $\vec{r}(t) = t^2\vec{i} + t\vec{j}$, we have $\vec{r}\,'(t) = 2t\vec{i} + \vec{j}$. Thus,

$$\int_C \vec{F} \cdot d\vec{r} = \int_0^1 \vec{F}(t^2, t) \cdot (2t\vec{i} + \vec{j})\,dt$$

$$= \int_0^1 [(3t^2 - t)\vec{i} + t^2\vec{j}] \cdot (2t\vec{i} + \vec{j})\,dt$$

$$= \int_0^1 (6t^3 - t^2)\,dt$$

$$= (\frac{3t^4}{2} - \frac{t^3}{3})\Big|_0^1$$

$$= \frac{3}{2} - \frac{1}{3} - (0 - 0) = \frac{7}{6}$$

17. The line integral is defined by chopping the curve C into little pieces, C_i, and forming the sum

$$\sum_{C_i} \vec{F} \cdot \Delta\vec{r}.$$

When the pieces are small, $\Delta\vec{r}$ is approximately tangent to C_i, and its magnitude is approximately equal to the length of the little piece of curve C_i. This means that $\vec{F}$ and $\Delta\vec{r}$ are almost parallel, the dot product is approximately equal to the product of their magnitudes, i.e.,

$$\vec{F} \cdot \Delta\vec{r} \approx m \cdot (\text{Length of } C_i).$$

When we sum all the dot products, we get

$$\sum_{C_i} \vec{F} \cdot \Delta\vec{r} \approx \sum_{C_i} m \cdot (\text{Length of } C_i)$$

$$= m \cdot \sum_{C_i} (\text{Length of } C_i)$$

$$= m \cdot (\text{Length of } C)$$

21. The integral corresponding to $A(t) = (t, t)$ is

$$\int_0^1 3t\,dt.$$

The integral corresponding to $D(t) = (e^t - 1, e^t - 1)$ is

$$3\int_0^{\ln 2} (e^{2t} - e^t)\,dt.$$

The substitution $s = e^t - 1$ has $ds = e^t\,dt$. Also $s = 0$ when $t = 0$ and $s = 1$ when $t = \ln 2$. Thus, substituting into the integral corresponding to $D(t)$ and using the fact that $e^{2t} = e^t \cdot e^t$ gives

$$3\int_0^{\ln 2} (e^{2t} - e^t)\,dt = 3\int_0^{\ln 2} (e^t - 1)e^t\,dt = \int_0^1 3s\,ds.$$

The integral on the right-hand side is the same as the integral corresponding to $A(t)$. Therefore we have

$$3\int_0^{\ln 2} (e^{2t} - e^t)\,dt = \int_0^1 3s\,ds = \int_0^1 3t\,dt.$$

Alternatively, the substitution $t = e^w - 1$ converts the integral corresponding to $A(t)$ into the integral corresponding to $B(t)$.

Solutions for Section 18.3

Exercises

1. The vector field $\vec{F}$ points radially outward, and so is everywhere perpendicular to A; thus, $\int_A \vec{F} \cdot d\vec{r} = 0$.

Along the first half of B, the terms $\vec{F} \cdot \Delta\vec{r}$ are negative; along the second half the terms $\vec{F} \cdot \Delta\vec{r}$ are positive. By symmetry the positive and negative contributions cancel out, giving a Riemann sum and a line integral of 0.

The line integral is also 0 along C, by cancellation. Here the values of $\vec{F}$ along the x-axis have the same magnitude as those along the y-axis. On the first half of C the path is traversed in the opposite direction to $\vec{F}$; on the second half of C the path is traversed in the same direction as $\vec{F}$. So the two halves cancel.

5. No. Suppose there were a function f such that grad $f = \vec{F}$. Then we would have

$$\frac{\partial f}{\partial x} = \frac{-z}{\sqrt{x^2+z^2}}.$$

Hence we would have

$$\frac{\partial^2 f}{\partial y \partial x} = \frac{\partial}{\partial y}(\frac{-z}{\sqrt{x^2+z^2}}) = 0.$$

In addition, since grad $f = \vec{F}$, we have that

$$\frac{\partial f}{\partial y} = \frac{y}{\sqrt{x^2+z^2}}.$$

Thus we also know that

$$\frac{\partial^2 f}{\partial x \partial y} = \frac{\partial}{\partial x}\left(\frac{y}{\sqrt{x^2+y^2}}\right) = -xy(x^2+z^2)^{-3/2}.$$

Notice that

$$\frac{\partial^2 f}{\partial y \partial x} \neq \frac{\partial^2 f}{\partial x \partial y}.$$

Since we expect $\frac{\partial^2 f}{\partial y \partial x} = \frac{\partial^2 f}{\partial x \partial y}$, we have got a contradiction. The only way out of this contradiction is to conclude there is no function f with grad $f = \vec{F}$. Thus $\vec{F}$ is not a gradient vector field.

9. Path independent

13. Since $\vec{F} = 2x\vec{i} - 4y\vec{j} + (2z-3)\vec{k} = \text{grad}(x^2 - 2y^2 + z^2 - 3z)$, the Fundamental Theorem of Line Integrals gives

$$\int_C \vec{F} \cdot d\vec{r} = (x^2 - 2y^2 + z^2 - 3z)\Big|_{(1,1,1)}^{(2,3,-1)} = (4 - 2\cdot 3^2 + (-1)^2 + 3) - (1^2 - 2\cdot 1^2 + 1^2 - 3) = -7.$$

Problems

17. Since $\vec{F} = \text{grad}\left(\dfrac{x^2+y^2}{2}\right)$, the line integral can be calculated using the Fundamental Theorem of Line Integrals:

$$\int_c \vec{F} \cdot d\vec{r} = \frac{x^2+y^2}{2}\Big|_{(0,0)}^{(3/\sqrt{2},3/\sqrt{2})} = \frac{9}{2}.$$

21. Since $\vec{F} = \text{grad} f$ is a gradient vector field, the Fundamental Theorem of Line Integrals give us

$$\int_C \vec{F} \cdot d\vec{r} = f(\text{end}) - f(\text{start}) = \left(x^2 + 2y^3 + 3z^4\right)\Big|_{(4,0,0)}^{(0,0,5)} = 3\cdot(5)^4 - 4^2 = 1859.$$

25. (a) Work done by the force is the line integral, so

$$\text{Work done against force} = -\int_C \vec{F}\cdot d\vec{r} = -\int_C(-mg\vec{k})\cdot d\vec{r}.$$

Since $\vec{r} = (\cos t)\vec{i} + (\sin t)\vec{j} + t\vec{k}$, we have $\vec{r}\,' = -(\sin t)\vec{i} + (\cos t)\vec{j} + \vec{k}$,

$$\begin{aligned}\text{Work done against force} &= \int_0^{2\pi} mg\vec{k}\cdot(-\sin t\vec{i} + \cos t\vec{j} + \vec{k})dt\\ &= \int_0^{2\pi} mg\,dt = 2\pi\,mg.\end{aligned}$$

(b) We know from physical principles that the force is conservative. (Because the work done depends only on the vertical distance moved, not on the path taken.) Alternatively, we see that

$$\vec{F} = -mg\vec{k} = \text{grad}(-mgz),$$

so $\vec{F}$ is a gradient field and therefore path independent, or conservative.

29. (a) The vector field ∇f is perpendicular to the level curves, in direction of increasing f. The length of ∇f is the rate of change of f in that direction. See Figure 18.5

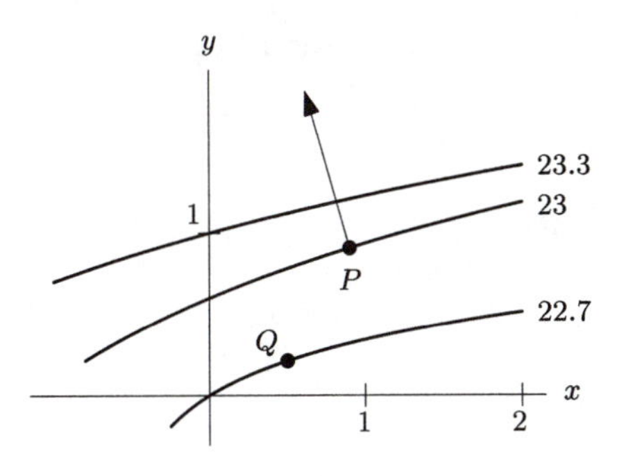

Figure 18.5

(b) Longer.

(c) Using the Fundamental Theorem of Calculus for Line Integrals, we have

$$\int_C \nabla f\cdot d\vec{r} = f(Q) - f(P) = 22.7 - 23 = -0.3.$$

Solutions for Section 18.4

Exercises

1. We know that

$$\frac{\partial f}{\partial x} = 2xy \quad\text{and}\quad \frac{\partial f}{\partial y} = x^2 + 8y^3$$

Now think of y as a constant in the equation for $\partial f/\partial x$ and integrate, giving

$$f(x,y) = x^2y + C(y).$$

Since the constant of integration may depend on y, it is written $C(y)$. Differentiating this expression for $f(x,y)$ with respect to y and using the fact that $\partial f/\partial y = x^2 + 8y^3$, we get

$$\frac{\partial f}{\partial y} = x^2 + C'(y) = x^2 + 8y^3.$$

Therefore

$$C'(y) = 8y^3 \quad \text{so} \quad C(y) = 2y^4 + K.$$

for some constant K. Thus,

$$f(x, y) = x^2y + 2y^4 + K.$$

5. The domain of the vector field $\vec{F} = (2xy^3 + y)\vec{i} + (3x^2y^2 + x)\vec{j}$ is the whole xy-plane. We apply the curl test:

$$\frac{\partial F_1}{\partial y} = 6xy^2 + 1 = \frac{\partial F_2}{\partial x}$$

so $\vec{F}$ is the gradient of a function f. In order to compute f we first integrate

$$\frac{\partial f}{\partial x} = 2xy^3 + y$$

with respect to x thinking of y as a constant. We get

$$f(x, y) = x^2y^3 + xy + C(y)$$

Differentiating with respect to y and using the fact that $\partial f/\partial y = 3x^2y^2 + x$ gives

$$\frac{\partial f}{\partial y} = 3x^2y^2 + x + C'(y) = 3x^2y^2 + x$$

Thus $C'(y) = 0$ so C is constant and

$$f(x, y) = x^2y^3 + xy + C.$$

9. By Green's Theorem, with R representing the interior of the circle,

$$\int_C \vec{F} \cdot d\vec{r} = \int_R \left(\frac{\partial}{\partial x}(-x) - \frac{\partial}{\partial y}(y) \right) dA = -2 \int_R dA$$
$$= -2 \cdot \text{ Area of circle } = -2\pi(1^2) = -2\pi.$$

Problems

13. Since the level curves must be perpendicular to the gradient vectors, if there were a contour diagram fitting this gradient field, it would have to look like Figure 18.6. However, this diagram could not be the contour diagram because the origin is on all contours. This means that $f(0, 0)$ would have to take on more than one value, which is impossible. At a point P other than the origin, we have the same problem. The values on the contours increase as you go counterclockwise around, since the gradient vector points in the direction of greatest increase of a function. But, starting at P, and going all the way around the origin, you would eventually get back to P again, and with a larger value of f, which is impossible.

An additional problem arises from the fact that the vectors in the original vector field are longer as you go away from the origin. This means that if there were a potential function f then $||\text{grad } f||$ would increase as you went away from the origin. This would mean that the level curves of f would get closer together as you go outward which does not happen in the contour diagram in Figure 18.6.

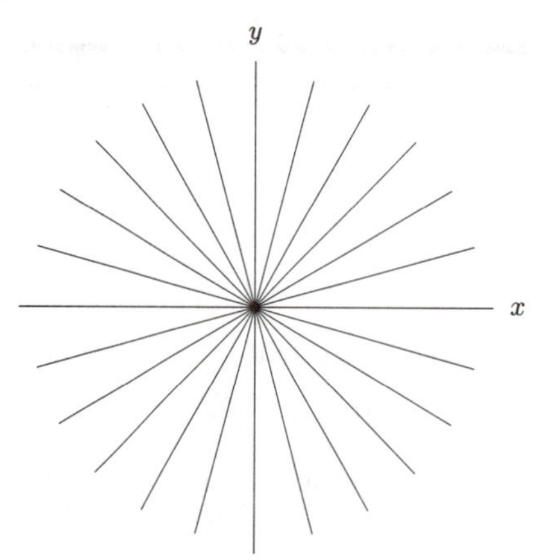

Figure 18.6

17. **(a)** C_1 is a line along the vertical axis; C_2 is a half circle from the positive y to the negative y-axis. See Figure 18.7.

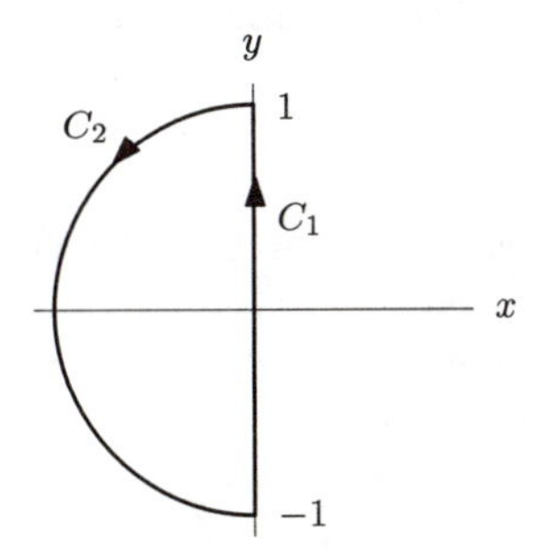

Figure 18.7

(b) Either use Green's Theorem or calculate directly. Using Green's Theorem, with R as the region inside C, we get

$$\int_{C_1+C_2} \vec{F}\cdot d\vec{r} = \int_R \left(\frac{\partial}{\partial x}(y) - \frac{\partial}{\partial y}(x+3y)\right)\, dA$$
$$= \int_R -3\, dA = -3(\text{Area of region}) = -3\frac{\pi\cdot 1^2}{2} = -\frac{3\pi}{2}.$$

21. Using $\vec{F} = x\vec{j} = a\cos^3 t$ and $\vec{r}\,'(t) = -3a\cos^2 t\sin t\vec{i} + 3a\sin^2 t\cos t\vec{j}$, we have

$$A = \int_C \vec{F}\cdot d\vec{r} = \int_0^{2\pi} (a\cos^3 t)(3a\sin^2 t\cos t)\, dt$$
$$= 3a^2\int_0^{2\pi} \cos^4 t\sin^2 t\, dt = 3a^2\int_0^{2\pi} \cos^2 t(\sin t\cos t)^2\, dt = 3a^2\int_0^{2\pi} \cos^2 t\frac{\sin^2 2t}{4}\, dt$$
$$= \frac{3a^2}{16}\int_0^{2\pi} (1+\cos 2t)(1-\cos 4t)\, dt$$
$$= \frac{3a^2}{16}\int_0^{2\pi} (1+\cos 2t - \cos 4t - \cos 2t\cos 4t)\, dt$$
$$= \frac{3a^2}{16}\int_0^{2\pi} (1+\cos 2t - \cos 4t - \frac{1}{2}\cos 6t - \frac{1}{2}\cos 2t)\, dt$$
$$= \frac{3a^2}{16}(t - \frac{1}{2}\sin 2t - \frac{1}{4}\sin 4t + \frac{1}{12}\sin 6t + \frac{1}{4}\sin 2t)\Big|_0^{2\pi} = \frac{3\pi a^2}{8}$$

For the last integral we use the trigonometric formula $\cos 2t\cos 4t = \frac{1}{2}(\cos 6t + \cos 2t)$. The hypocycloid is shown in Figure 18.8.

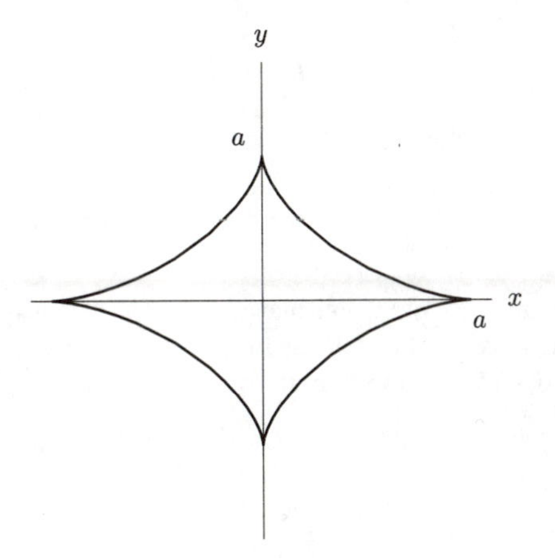

Figure 18.8: $x^{2/3} + y^{2/3} = a^{2/3}$

Solutions for Chapter 18 Review

Exercises

1. On the top half of the circle, the angle between the vector field and the curve is less than $90°$, so the line integral is positive. On the bottom half of the circle, the angle is more than $90°$, so the line integral is negative. However the magnitude of the vector field is larger on the top half of the curve, so the positive contribution to the line integral is larger than the negative. Thus the line integral $\int_C \vec{F} \cdot d\vec{r}$ is positive.

5. We parameterize the path C by (t,t,t), for $0 \le t \le 1$. Then

$$\int_C \vec{F} \cdot d\vec{r} = \int_0^1 \vec{F}(t,t,t) \cdot (\vec{i} + \vec{j} + \vec{k})dt = \int_0^1 (t\vec{i} + 3t\vec{j} - t^2\vec{k}) \cdot (\vec{i} + \vec{j} + \vec{k})dt$$

$$= \int_0^1 (4t - t^2)dt = \left(2t^2 - \frac{t^3}{3}\right)\Big|_0^1 = \frac{5}{3}.$$

9. Since $\vec{F} = \text{grad}\left(\frac{x^2}{2} + \frac{y^2}{2} + \frac{z^2}{2}\right)$, the Fundamental Theorem of Line Integrals gives

$$\int_C \vec{F} \cdot d\vec{r} = \left(\frac{x^2}{2} + \frac{y^2}{2} + \frac{z^2}{2}\right)\Bigg|_{(2,3,0)}^{(0,0,7)} = \frac{7^2}{2} - \left(\frac{2^2}{2} + \frac{3^2}{2}\right) = 18.$$

13. The domain is all 3-space. Since $F_1 = y$, $F_2 = x$,

$$\text{curl}\, y\vec{i} + x\vec{j} = \left(\frac{\partial F_3}{\partial y} - \frac{\partial F_2}{\partial z}\right)\vec{i} + \left(\frac{\partial F_1}{\partial z} - \frac{\partial F_3}{\partial x}\right)\vec{j} + \left(\frac{\partial F_2}{\partial x} - \frac{\partial F_1}{\partial y}\right)\vec{k} = \vec{0},$$

so $\vec{F}$ is path-independent

Problems

17. **(a)** Since $\vec{F} = x\vec{i} + y\vec{j} = \text{grad}\left(\frac{x^2+y^2}{2}\right)$, we know that $\vec{F}$ is a gradient vector field. Thus, by the Fundamental Theorem of Line Integrals,

$$\int_{OA} \vec{F} \cdot d\vec{r} = \frac{x^2+y^2}{2}\Bigg|_{(0,0)}^{(3,0)} = \frac{9}{2}.$$

(b) We know that $\vec{F}$ is path independent. If C is the closed curve consisting of the line in part (a) followed by the two-part curve in part (b), then

$$\int_C \vec{F} \cdot d\vec{r} = 0.$$

Thus, if ABO is the two-part curve of part (b) and OA is the line in part (a),

$$\int_{ABO} \vec{F} \cdot d\vec{r} = -\int_{OA} \vec{F} \cdot d\vec{r} = -\frac{9}{2}.$$

21. Yes, the line integral over C_1 is the negative of the line integral over C_2. One way to see this is to observe that the vector field $x\vec{i} + y\vec{j}$ is symmetric in the y-axis and that C_1 and C_2 are reflections in the y axis (except for orientation). See Figure 18.9. Since the orientation of C_2 is the reverse of the orientation of a mirror image of C_1, the two line integrals are opposite in sign.

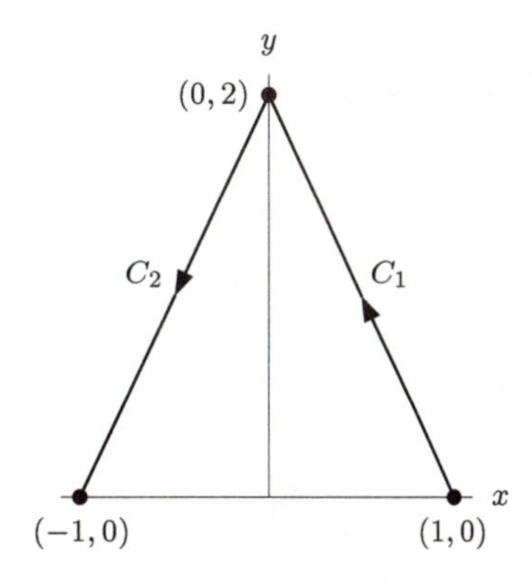

Figure 18.9

25. We'll assume that the rod is positioned along the z-axis, and look at the magnetic field $\vec{B}$ in the xy-plane. If C is a circle of radius r in the plane, centered at the origin, then we are told that the magnetic field is tangent to the circle and has constant magnitude $\|\vec{B}\|$. We divide the curve C into little pieces C_i and then we sum $\vec{B}\cdot\Delta\vec{r}$ computed on each piece C_i. But $\Delta\vec{r}$ points nearly in the same direction as $\vec{B}$, that is, tangent to C, and has magnitude nearly equal to the length of C_i. So the dot product is nearly equal to $\|\vec{B}\|\times$ length of C_i. When all of these dot products are summed and the limit is taken as $\|\Delta\vec{r}\|\to 0$, we get

$$\int_C \vec{B}\cdot d\vec{r} = \|\vec{B}\| \times \text{length of } C = \|\vec{B}\| \times 2\pi r$$

Now Ampère's Law also tells us that

$$\int_C \vec{B}\cdot d\vec{r} = kI$$

Setting these expressions for the line integral equal to each other and solving for $\|\vec{B}\|$ gives $kI = 2\pi r\|\vec{B}\|$, so

$$\|\vec{B}\| = \frac{kI}{2\pi r}.$$

CAS Challenge Problems

29. **(a)** We parameterize C_a by $\vec{r}(t) = a\cos t\vec{i} + a\sin t\vec{j}$. Then, using a CAS, we find

$$\begin{aligned}\int_{C_a} \vec{F}(\vec{r}(t))\cdot\vec{r}\,'(t)\,dt &= \int_0^{2\pi} a\cos t\left(2a\cos t - \frac{a^3\cos t^3}{3} + a^3\cos t\sin t^2\right)\\ &\quad -a\sin t\left(-(a\sin t) + \frac{2a^3\sin t^3}{3}\right)\,dt\\ &= -\frac{\pi}{2}(-6a^2 + a^4)\end{aligned}$$

The derivative of the expression on the right with respect to a is $-(2\pi)(-3a + a^3)$, which is zero at $a = 0, \pm\sqrt{3}$. Checking at $a = 0$ and as $a\to\infty$, we find the maximum is at $a = \sqrt{3}$.

(b) We have

$$\frac{\partial F_2}{\partial x} - \frac{\partial F_1}{\partial y} = (2 - x^2 + y^2) - (-1 + 2y^2) = 3 - x^2 - y^2.$$

So, by Green's theorem,

$$\int_{C_a} \vec{F}\cdot d\vec{r} = \int\int_{D_a} (3 - x^2 - y^2)\,dA,$$

where D_a is the disc of radius a centered at the origin. The integrand is positive for $x^2 + y^2 < 3$, so it is positive inside the disc of radius $\sqrt{3}$ and negative outside it. Thus the integral has its maximum value when $a = \sqrt{3}$.

CHECK YOUR UNDERSTANDING

1. A path-independent vector field must have zero circulation around all closed paths. Consider a vector field like $\vec{F}(x, y) = |x|\vec{j}$, shown in Figure 18.10.

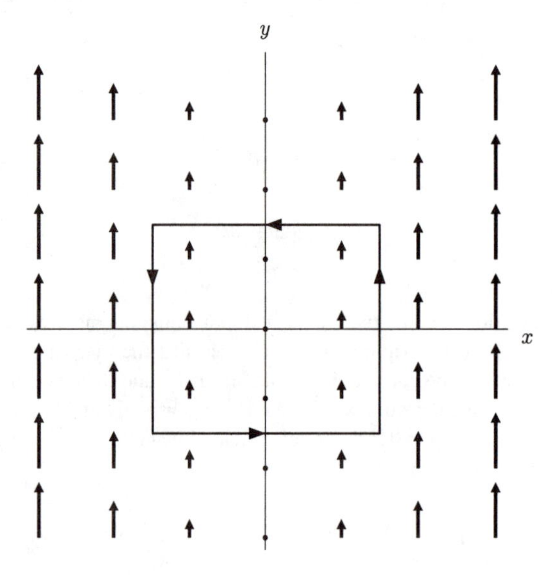

Figure 18.10

A rectangular path that is symmetric about the y-axis will have zero circulation: on the horizontal sides, the field is perpendicular, so the line integral is zero. The line integrals on the vertical sides are equal in magnitude and opposite in sign, so they cancel out, giving a line integral of zero. However, this field is not path-independent, because it is possible to find two paths with the same endpoints but different values of the line integral of $\vec{F}$. For example, consider the two points $(0, 0)$ and $(0, 1)$. The path C_1 in Figure 18.11 along the y axis gives zero for the line integral, because the field is 0 along the y axis, whereas a path like C_2 will have a nonzero line integral. Thus the line integral depends on the path between the points, so $\vec{F}$ is not path-independent.

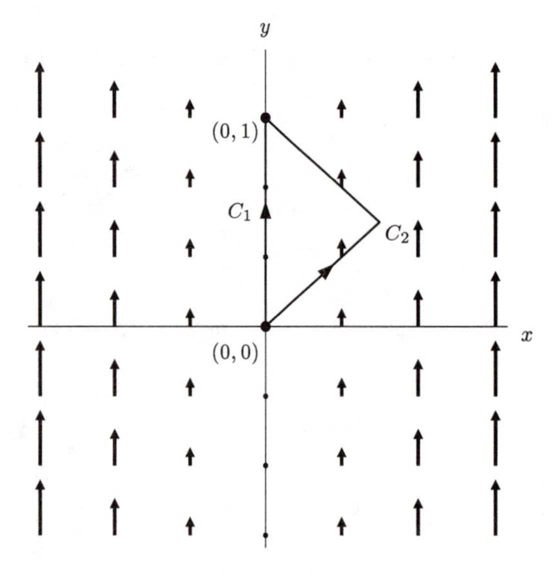

Figure 18.11

5. False. Because $\vec{F} \cdot \Delta\vec{r}$ is a scalar quantity, $\int_C \vec{F} \cdot d\vec{r}$ is also a scalar quantity.

9. True. The line integral is the limit of a sum of dot products, hence is a scalar.

13. False. All of the dot products $\vec{F}(\vec{r}_i) \cdot \Delta\vec{r}_i$ in this line integral are zero, since the vector field (the constant $\vec{i}$) points perpendicular to $\Delta\vec{r}_i$.

17. False. The line integrals of many vector fields (so called *path independent* or *conservative fields*) are zero around closed curves, but this is not true of all fields. For example, a vector field that is flowing in the same direction as the curve C all along the curve has a positive line integral. A specific example is given by $\vec{F} = -y\vec{i} + x\vec{j}$, where C is the unit circle centered at the origin, oriented counterclockwise.

21. False. As a counterexample, consider the unit circle C, centered at the origin, oriented counterclockwise and the vector field $\vec{F} = -y\vec{i} + x\vec{j}$. The vector field is always tangent to the circle, and in the same direction as C, so the line integral is positive.

25. True. The curves C_1 and C_2 are the same (they follow the graph of $y = x^2$ between $(0,0)$ and $(2,4)$), except that their orientations are opposite.

29. True. The construction at the end of Section 18.3 shows how to make a potential function from a path-independent vector field.

33. False. For example, take $\vec{F} = y\vec{i}$. By symmetry, the line integral of $\vec{F}$ over any circle centered at the origin is zero. But the curve consisting of the upper semicircle connecting $(-a,0)$ to $(a,0)$ has a positive line integral, while the line connecting these points along the x-axis has a zero line integral, so the field cannot be path-independent.

37. False. As a counterexample, consider $\vec{F} = x\vec{j}$ and $\vec{G} = y\vec{i}$. Then both of these are path-dependent (they each have nonzero curl), but the curl of $\vec{F} + \vec{G} = y\vec{i} + x\vec{j}$ is zero everywhere, so $\vec{F} + \vec{G}$ is path-independent.

41. False. As a counterexample, consider the vector field $\vec{F} = 2x\vec{i}$, which is path-independent, since it is the gradient of $f(x,y) = x^2$. Multiplying $\vec{F}$ by the function $h(x,y) = y$ gives the field $y\vec{F} = 2xy\vec{i}$. The curl of this vector field is $-2x \neq 0$, so $y\vec{F}$ is path-dependent.

CHAPTER NINETEEN

Solutions for Section 19.1

Exercises

1. **(a)** The flux is positive, since $\vec{F}$ points in direction of positive x-axis, the same direction as the normal vector.
(b) The flux is negative, since below the xy-plane $\vec{F}$ points towards negative x-axis, which is opposite the orientation of the surface.
(c) The flux is zero. Since $\vec{F}$ has only an x-component, there is no flow across the surface.
(d) The flux is zero. Since $\vec{F}$ has only an x-component, there is no flow across the surface.
(e) The flux is zero. Since $\vec{F}$ has only an x-component, there is no flow across the surface.

5. $\vec{v} \cdot \vec{A} = (2\vec{i} + 3\vec{j} + 5\vec{k}) \cdot 2\vec{i} = 4.$

9. The square has area 16, so its area vector is $16\vec{j}$. Since $\vec{F} = 5\vec{j}$ on the square,

$$\text{Flux} = 5\vec{j} \cdot 16\vec{j} = 80.$$

13. The disk has area 25π, so its area vector is $25\pi\vec{j}$. Thus

$$\text{Flux} = (2\vec{i} + 3\vec{j}) \cdot 25\pi\vec{j} = 75\pi.$$

17. Since the disk is in the xy-plane and oriented upward, $d\vec{A} = \vec{k}\, dxdy$ and

$$\int_{\text{Disk}} \vec{F} \cdot d\vec{A} = \int_{\text{Disk}} (x^2 + y^2)\vec{k} \cdot \vec{k}\, dxdy = \int_{\text{Disk}} (x^2 + y^2)\, dxdy.$$

Using polar coordinates

$$\int_{\text{Disk}} \vec{F} \cdot d\vec{A} = \int_0^{2\pi} \int_0^3 r^2 \cdot r\, dr d\theta = 2\pi \left. \frac{r^4}{4} \right|_0^3 = \frac{81\pi}{2}.$$

Problems

21. On the curved sides of the cylinder, the $\vec{k}$ component of $\vec{F}$ does not contribute to the flux. Since the $\vec{i}$ and $\vec{j}$ components are constant, these components contribute 0 to the flux on the entire cylinder. Therefore the only nonzero contribution to the flux results from the $\vec{k}$ component through the top, where $z = 2$ and $d\vec{A} = \vec{k}\, dA$, and from the $\vec{k}$ component through the bottom, where $z = -2$ and $d\vec{A} = -\vec{k}\, dA$:

$$\begin{aligned} \text{Flux} &= \int_{\text{Top}} \vec{F} \cdot d\vec{A} + \int_{\text{Bottom}} \vec{F} \cdot d\vec{A} \\ &= \int_{\text{Top}} 2\vec{k} \cdot \vec{k}\, dA + \int_{\text{Bottom}} (-2\vec{k}) \cdot (-\vec{k}\, dA) \\ &= 4 \int_{\text{Top}} dA = 4 \cdot \text{Area of top} = 4 \cdot \pi(3^2) = 36\pi. \end{aligned}$$

25. **(a)** At the north pole, the area vector of the plate is upward (away from the center of the earth), and so is in the opposite direction to the magnetic field. Thus the magnetic flux is negative.
(b) At the south pole, the area vector of the plate is again away from the center of the earth (because that is upward in the southern hemisphere), and so is in the same direction as the magnetic field. Thus, the magnetic flux is positive.
(c) At the equator the magnetic field is parallel to the plate, so the flux is zero.

29. **(a)** Figure 19.1 shows the electric field $\vec{E}$. Note that $\vec{E}$ points radially outward from the z-axis.

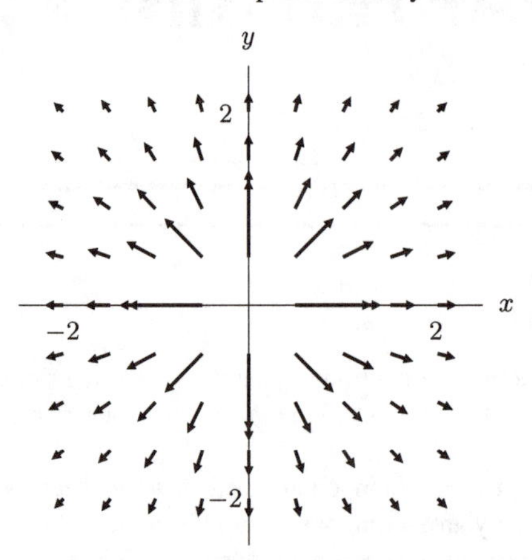

Figure 19.1: The electric field in the xy-plane due to a line of positive charge uniformly distributed along the z-axis: $\vec{E}(x, y, 0) = 2\lambda \dfrac{x\vec{i} + y\vec{j}}{x^2 + y^2}$

(b) On the cylinder $x^2 + y^2 = R^2$, the electric field $\vec{E}$ points in the same direction as the outward normal $\vec{n}$, and

$$\|\vec{E}\| = \frac{2\lambda}{R^2}\|x\vec{i} + y\vec{j}\| = \frac{2\lambda}{R}.$$

So

$$\begin{aligned}\int_S \vec{E} \cdot d\vec{A} &= \int_S \vec{E} \cdot \vec{n}\, dA = \int_S \|\vec{E}\|\, dA = \int_S \frac{2\lambda}{R}\, dA \\ &= \frac{2\lambda}{R}\int_S dA = \frac{2\lambda}{R} \cdot \text{Area of } S = \frac{2\lambda}{R} \cdot 2\pi Rh = 4\pi\lambda h,\end{aligned}$$

which is positive, as we expected.

33. **(a)** From Newton's law of cooling, we know that the temperature gradient will be proportional to the heat flow. If the constant of proportionality is k then we have the equation $\vec{F} = k \operatorname{grad} T$. Since $\operatorname{grad} T$ points in the direction of increasing T, but heat flows towards lower temperatures, the constant k must be negative.

(b) This form of Newton's law of cooling is saying that heat will be flowing in the direction in which temperature is decreasing most rapidly, in other words, in the direction exactly opposite to $\operatorname{grad} T$. This agrees with our intuition which tells us that a difference in temperature causes heat to flow from the higher temperature to the lower temperature, and the rate at which it flows depends on the temperature gradient.

(c) The rate of heat loss from W is given by the flux of the heat flow vector field through the surface of the body. Thus,

$$\begin{array}{c}\text{Rate of heat}\\ \text{loss from } W\end{array} = \begin{array}{c}\text{Flux of } \vec{F}\\ \text{out of } S\end{array} = \int_S \vec{F} \cdot d\vec{A} = k\int_S (\operatorname{grad} T) \cdot d\vec{A}$$

Solutions for Section 19.2

Exercises

1. Using $z = f(x, y) = x + y$, we have $d\vec{A} = (-\vec{i} - \vec{j} + \vec{k})\, dx\, dy$. As S is oriented upward, we have

$$\begin{aligned}\int_S \vec{F} \cdot d\vec{A} &= \int_0^3\int_0^2 ((x - y)\vec{i} + (x + y)\vec{j} + 3x\vec{k}) \cdot (-\vec{i} - \vec{j} + \vec{k})\, dxdy \\ &= \int_0^3\int_0^2 (-x + y - x - y + 3x)\, dxdy = \int_0^3\int_0^2 x\, dxdy = 6.\end{aligned}$$

5.

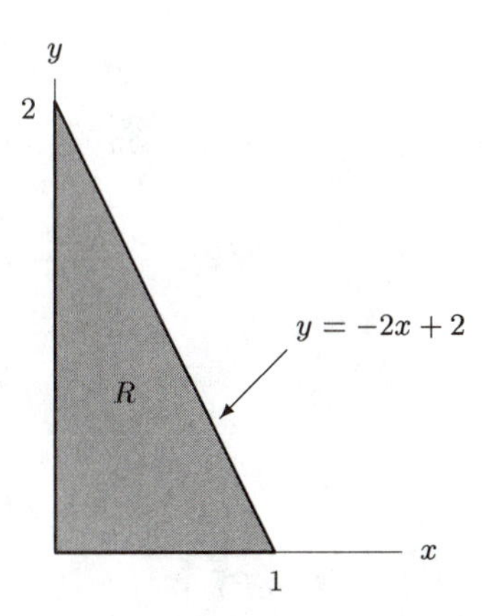

Figure 19.2

Writing the surface S as $z = f(x, y) = -2x - 4y + 1$, we have

$$d\vec{A} = (-f_x\vec{i} - f_y\vec{j} + \vec{k})dxdy.$$

With R as shown in Figure 19.2, we have

$$\begin{aligned}
\int_S \vec{F} \cdot d\vec{A} &= \int_R \vec{F}(x, y, f(x, y)) \cdot (-f_x\vec{i} - f_y\vec{j} + \vec{k})\, dxdy \\
&= \int_R (3x\vec{i} + y\vec{j} + (-2x - 4y + 1)\vec{k}) \cdot (2\vec{i} + 4\vec{j} + \vec{k})\, dxdy \\
&= \int_R (4x + 1)\, dxdy = \int_0^1 \int_0^{-2x+2} (4x + 1)\, dydx \\
&= \int_0^1 (4x + 1)(-2x + 2)\, dx \\
&= \int_0^1 (-8x^2 + 6x + 2)\, dx = \left(-\frac{8x^3}{3} + 3x^2 + 2x\right)\Big|_0^1 = \frac{7}{3}.
\end{aligned}$$

9. We have $0 \le z \le 6$ so $0 \le x^2 + y^2 \le 36$. Let R be the disk of radius 6 in the xy-plane centered at the origin. Because of the cone's point, the flux integral is improper; however, it does converge. We have

$$\begin{aligned}
\int_S \vec{F} \cdot d\vec{A} &= \int_R \vec{F}(x, y, f(x, y)) \cdot (-f_x\vec{i} - f_y\vec{j} + \vec{k})\, dxdy \\
&= \int_R (-x\sqrt{x^2 + y^2}\vec{i} - y\sqrt{x^2 + y^2}\vec{j} + (x^2 + y^2)\vec{k}) \\
&\qquad \cdot \left(-\frac{x}{\sqrt{x^2 + y^2}}\vec{i} - \frac{y}{\sqrt{x^2 + y^2}}\vec{j} + \vec{k}\right) dxdy \\
&= \int_R 2(x^2 + y^2)\, dxdy \\
&= 2\int_0^6 \int_0^{2\pi} r^3\, d\theta dr \\
&= 4\pi \int_0^6 r^3\, dr = 1296\pi.
\end{aligned}$$

13. Since the radius of the cylinder is 1, using cylindrical coordinates we have

$$d\vec{A} = (\cos\theta\vec{i} + \sin\theta\vec{j})d\theta dz.$$

Thus,

$$\begin{aligned}\int_S \vec{F} \cdot d\vec{A} &= \int_0^6 \int_0^{2\pi} (\cos\theta\vec{i} + \sin\theta\vec{j}) \cdot (\cos\theta\vec{i} + \sin\theta\vec{j})\, d\theta\, dz \\ &= \int_0^6 \int_0^{2\pi} 1\, d\theta\, dz = 12\pi.\end{aligned}$$

Problems

17. On the disk, $z = 0$ and $d\vec{A} = \vec{k}\, dx\, dy$, so

$$\begin{aligned}\int_S \vec{F} \cdot d\vec{A} &= \int_{x^2+y^2\le 1} (xze^{yz}\vec{i} + x\vec{j} + (5 + x^2 + y^2)\vec{k}) \cdot \vec{k}\, dx\, dy \\ &= \int_{x^2+y^2\le 1} (5 + x^2 + y^2)\, dx\, dy = \int_0^{2\pi} \int_0^1 (5 + r^2) r\, dr\, d\theta \\ &= 2\pi \left(\frac{5r^2}{2} + \frac{r^4}{4} \right) \Bigg|_0^1 = \frac{11\pi}{2}.\end{aligned}$$

Solutions for Section 19.3

Exercises

1. Since S is given by

$$\vec{r}(s,t) = (s+t)\vec{i} + (s-t)\vec{j} + (s^2+t^2)\vec{k},$$

we have

$$\frac{\partial \vec{r}}{\partial s} = \vec{i} + \vec{j} + 2s\vec{k} \quad \text{and} \quad \frac{\partial \vec{r}}{\partial t} = \vec{i} - \vec{j} + 2t\vec{k},$$

and

$$\frac{\partial \vec{r}}{\partial s} \times \frac{\partial \vec{r}}{\partial t} = \begin{vmatrix} \vec{i} & \vec{j} & \vec{k} \\ 1 & 1 & 2s \\ 1 & -1 & 2t \end{vmatrix} = (2s+2t)\vec{i} + (2s-2t)\vec{j} - 2\vec{k}.$$

Since the $\vec{i}$ component of this vector is positive for $0 < s < 1, 0 < t < 1$, it points away from the z-axis, and so has the opposite orientation to the one specified. Thus, we use

$$d\vec{A} = -\frac{\partial \vec{r}}{\partial s} \times \frac{\partial \vec{r}}{\partial t}\, ds\, dt,$$

and so we have

$$\begin{aligned}\int_S \vec{F} \cdot d\vec{A} &= -\int_0^1 \int_0^1 (s^2+t^2)\vec{k} \cdot \left((2s+2t)\vec{i} + (2s-2t)\vec{j} - 2\vec{k}\right) ds\, dt \\ &= 2\int_0^1 \int_0^1 (s^2+t^2)\, ds\, dt = 2\int_0^1 \left(\frac{s^3}{3} + st^2\right)\Bigg|_{s=0}^{s=1} dt \\ &= 2\int_0^1 (\frac{1}{3} + t^2)\, dt = 2(\frac{1}{3}t + \frac{t^3}{3})\Bigg|_0^1 = 2(\frac{1}{3} + \frac{1}{3}) = \frac{4}{3}.\end{aligned}$$

5. Using cylindrical coordinates, we see that the surface S is parameterized by

$$\vec{r}(r,\theta) = r\cos\theta\vec{i} + r\sin\theta\vec{j} + r\vec{k}.$$

We have

$$\frac{\partial\vec{r}}{\partial r} \times \frac{\partial\vec{r}}{\partial\theta} = \begin{vmatrix} \vec{i} & \vec{j} & \vec{k} \\ \cos\theta & \sin\theta & 1 \\ -r\sin\theta & r\cos\theta & 0 \end{vmatrix} = -r\cos\theta\vec{i} - r\sin\theta\vec{j} + r\vec{k}.$$

Since the vector $\partial\vec{r}/\partial r \times \partial\vec{r}/\partial\theta$ points upward, in the direction opposite to the specified orientation, we use $d\vec{A} = -(\partial\vec{r}/\partial r \times \partial\vec{r}/\partial\theta)\,dr\,d\theta$. Hence

$$\begin{aligned}
\int_S \vec{F}\cdot d\vec{A} &= -\int_0^{2\pi}\int_0^R (r^5\cos^2\theta\sin^2\theta\vec{k})\cdot(-r\cos\theta\vec{i} - r\sin\theta\vec{j} + r\vec{k})\,dr\,d\theta \\
&= -\int_0^{2\pi}\int_0^R r^6\cos^2\theta\sin^2\theta\,dr\,d\theta \\
&= -\frac{R^7}{7}\int_0^{2\pi}\sin^2\theta\cos^2\theta\,d\theta \\
&= -\frac{R^7}{7}\int_0^{2\pi}\sin^2\theta(1-\sin^2\theta)\,d\theta \\
&= -\frac{R^7}{7}\int_0^{2\pi}(\sin^2\theta - \sin^4\theta)\,d\theta \\
&= -(\frac{R^7}{7})(\frac{\pi}{4}) = \frac{-\pi}{28}R^7.
\end{aligned}$$

The cone is not differentiable at the point $(0,0)$. However the flux integral, which is improper, converges.

Problems

9. If S is the part of the graph of $z = f(x,y)$ lying over a region R in the xy-plane, then S is parameterized by

$$\vec{r}(x,y) = x\vec{i} + y\vec{j} + f(x,y)\vec{k}, \qquad (x,y) \text{ in } R.$$

So

$$\frac{\partial\vec{r}}{\partial x} \times \frac{\partial\vec{r}}{\partial y} = (\vec{i} + f_x\vec{k}) \times (\vec{j} + f_y\vec{k}) = -f_x\vec{i} - f_y\vec{j} + \vec{k}.$$

Since the $\vec{k}$ component is positive, this points upward, so if S is oriented upward

$$d\vec{A} = (-f_x\vec{i} - f_y\vec{j} + \vec{k})\,dx\,dy$$

and therefore we have the expression for the flux integral obtained on page 874:

$$\int_S \vec{F}\cdot d\vec{A} = \int_R \vec{F}(x,y,f(x,y))\cdot(-f_x\vec{i} - f_y\vec{k} + \vec{k})\,dx\,dy.$$

Solutions for Chapter 19 Review

Exercises

1. Since $\vec{G}$ is constant, the net flux through the sphere is 0, so

$$\int_S \vec{G}\cdot d\vec{A} = 0.$$

5. Since the vector field is everywhere perpendicular to the surface of the sphere, and $||\vec{F}|| = \pi$ on the surface, we have

$$\int_S \vec{F} \cdot d\vec{A} = ||\vec{F}|| \cdot \text{Area of sphere} = \pi \cdot 4\pi(\pi)^2 = 4\pi^4.$$

9. Only the $\vec{k}$ component contributes to the flux. In the plane $z = 4$, we have $\vec{F} = 2\vec{i} + 3\vec{j} + 4\vec{k}$. On the square $d\vec{A} = \vec{k}\, dA$, so we have

$$\text{Flux} = \int \vec{F} \cdot d\vec{A} = 4\vec{k} \cdot (\vec{k} \text{ Area of square}) = 4(5^2) = 100.$$

13. We have $d\vec{A} = \vec{k}\, dA$, so

$$\int_S \vec{F} \cdot d\vec{A} = \int_S (z\vec{i} + y\vec{j} + 2x\vec{k}) \cdot \vec{k}\, dA = \int_S 2x\, dA$$
$$= \int_0^3 \int_0^2 2x\, dx dy = 12.$$

17. There is no flux through the base or top of the cylinder because the vector field is parallel to these faces. For the curved surface, consider a small patch with area $\Delta\vec{A}$. The vector field is pointing radially outward from the z-axis and so is parallel to $\Delta\vec{A}$. Since $||\vec{F}|| = \sqrt{x^2 + y^2} = 2$ on the curved surface of the cylinder, we have $\vec{F} \cdot \Delta\vec{A} = ||\vec{F}||||\Delta\vec{A}|| = 2\Delta A$. Replacing ΔA with dA, we get

$$\int_S \vec{F} \cdot d\vec{A} = \int_{\substack{\text{Curved}\\\text{surface}}} 2\, dA = 2(\text{Area of curved surface}) = 2(2\pi \cdot 2 \cdot 3) = 24\pi.$$

Problems

21. The vector field $\vec{D}$ has constant magnitude on S, equal to $Q/4\pi R^2$, and points radially outward, so

$$\int_S \vec{D} \cdot d\vec{A} = \frac{Q}{4\pi R^2} \cdot 4\pi R^2 = Q.$$

25. **(a)** Consider two opposite faces of the cube, S_1 and S_2. The corresponding area vectors are $\vec{A}_1 = 4\vec{i}$ and $\vec{A}_2 = -4\vec{i}$ (since the side of the cube has length 2). Since $\vec{E}$ is constant, we find the flux by taking the dot product, giving

$$\text{Flux through } S_1 = \vec{E} \cdot \vec{A}_1 = (a\vec{i} + b\vec{j} + c\vec{k}) \cdot 4\vec{i} = 4a.$$
$$\text{Flux through } S_2 = \vec{E} \cdot \vec{A}_2 = (a\vec{i} + b\vec{j} + c\vec{k}) \cdot (-4\vec{i}) = -4a.$$

Thus the fluxes through S_1 and S_2 cancel. Arguing similarly, we conclude that, for any pair of opposite faces, the sum of the fluxes of $\vec{E}$ through these faces is zero. Hence, by addition, $\int_S \vec{E} \cdot d\vec{A} = 0$.

(b) The basic idea is the same as in part (a), except that we now need to use Riemann sums. First divide S into two hemispheres H_1 and H_2 by the equator C located in a plane perpendicular to $\vec{E}$. For a tiny patch S_1 in the hemisphere H_1, consider the patch S_2 in the opposite hemisphere which is symmetric to S_1 with respect to the center O of the sphere. The area vectors $\Delta\vec{A}_1$ and $\Delta\vec{A}_2$ satisfy $\Delta\vec{A}_2 = -\Delta\vec{A}_1$, so if we consider S_1 and S_2 to be approximately flat, then $\vec{E} \cdot \Delta\vec{A}_1 = -\vec{E} \cdot \Delta\vec{A}_2$. By decomposing H_1 and H_2 into small patches as above and using Riemann sums, we get

$$\int_{H_1} \vec{E} \cdot d\vec{A} = -\int_{H_2} \vec{E} \cdot d\vec{A}, \quad \text{so} \quad \int_S \vec{E} \cdot d\vec{A} = 0.$$

(c) The reasoning in part (b) can be used to prove that the flux of $\vec{E}$ through any surface with a center of symmetry is zero. For instance, in the case of the cylinder, cut it in half with a plane $z = 1$ and denote the two halves by H_1 and H_2. Just as before, take patches in H_1 and H_2 with $\Delta A_1 = -\Delta A_2$, so that $\vec{E} \cdot \Delta A_1 = -\vec{E} \cdot \Delta\vec{A}_2$. Thus, we get

$$\int_{H_1} \vec{E} \cdot d\vec{A} = -\int_{H_2} \vec{E} \cdot d\vec{A},$$

which shows that

$$\int_S \vec{E} \cdot d\vec{A} = 0.$$

CAS Challenge Problems

29. **(a)** When $x > 0$, the vector $x\vec{i}$ points in the positive x-direction, and when $x < 0$ it points in the negative x-direction. Thus it always points from the inside of the ellipsoid to the outside, so we expect the flux integral to be positive. The upper half of the ellipsoid is the graph of $z = f(x,y) = \frac{1}{\sqrt{2}}(1 - x^2 - y^2)$, so the flux integral is

$$\begin{aligned}\int_S \vec{F} \cdot d\vec{A} &= \int_{-1/2}^{1/2}\int_{-1/2}^{1/2} x\vec{i} \cdot (-f_x\vec{i} - f_y\vec{j} + \vec{k})\,dxdy \\ &= \int_{-1/2}^{1/2}\int_{-1/2}^{1/2} (-xf_x)\,dxdy = \int_{-1/2}^{1/2}\int_{-1/2}^{1/2} \frac{x^2}{\sqrt{1-x^2-y^2}}\,dxdy \\ &= \frac{-\sqrt{2} + 11\arcsin(\frac{1}{\sqrt{3}}) + 10\arctan(\frac{1}{\sqrt{2}}) - 8\arctan(\frac{5}{\sqrt{2}})}{12} = 0.0958.\end{aligned}$$

Different CASs may give the answer in different forms. Note that we could have predicted the integral was positive without evaluating it, since the integrand is positive everywhere in the region of integration.

(b) For $x > -1$, the quantity $x + 1$ is positive, so the vector field $(x+1)\vec{i}$ always points in the direction of the positive x-axis. It is pointing into the ellipsoid when $x < 0$ and out of it when $x > 0$. However, its magnitude is smaller when $-1/2 < x < 0$ than it is when $0 < x < 1/2$, so the net flux out of the ellipsoid should be positive. The flux integral is

$$\begin{aligned}\int_S \vec{F} \cdot d\vec{A} &= \int_{-1/2}^{1/2}\int_{-1/2}^{1/2} (x+1)\vec{i} \cdot (-f_x\vec{i} - f_y\vec{j} + \vec{k})\,dxdy \\ &= \int_{-1/2}^{1/2}\int_{-1/2}^{1/2} -(x+1)f_x\,dxdy = \int_{-1/2}^{1/2}\int_{-1/2}^{1/2} \frac{x(1+x)}{\sqrt{1-x^2-y^2}}\,dxdy \\ &= \frac{\sqrt{2} - 11\arcsin(\frac{1}{\sqrt{3}}) - 10\arctan(\frac{1}{\sqrt{2}}) + 8\arctan(\frac{5}{\sqrt{2}})}{12} = 0.0958\end{aligned}$$

The answer is the same as in part (a). This makes sense because the difference between the integrals in parts (a) and (b) is the integral of $\int_{-1/2}^{1/2}\int_{-1/2}^{1/2}(x/\sqrt{1-x^2-y^2})\,dxdy$, which is zero because the integrand is odd with respect to x.

(c) This integral should be positive for the same reason as in part (a). The vector field $y\vec{j}$ points in the positive y-direction when $y > 0$ and in the negative y-direction when $y < 0$, thus it always points out of the ellipsoid. Evaluating the integral we get

$$\begin{aligned}\int_S \vec{F} \cdot d\vec{A} &= \int_{-1/2}^{1/2}\int_{-1/2}^{1/2} y\vec{j} \cdot (-f_x\vec{i} - f_y\vec{j} + \vec{k})\,dxdy \\ &= \int_{-1/2}^{1/2}\int_{-1/2}^{1/2} (-yf_y)\,dxdy = \int_{-1/2}^{1/2}\int_{-1/2}^{1/2} \frac{y^2}{\sqrt{1-x^2-y^2}}\,dxdy \\ &= \frac{\sqrt{2} - 2\arcsin(\frac{1}{\sqrt{3}}) - 19\arctan(\frac{1}{\sqrt{2}}) + 8\arctan(\frac{5}{\sqrt{2}})}{12} = 0.0958.\end{aligned}$$

The symbolic answer appears different but has the same numerical value as in parts (a) and (b). In fact the answer is the same because the integral here is the same as in part (a) except that the roles of x and y have been exchanged. Different CASs may give different symbolic forms.

CHECK YOUR UNDERSTANDING

1. True. By definition, the flux integral is the limit of a sum of dot products, hence is a scalar.

5. True. The flow of this field is in the same direction as the orientation of the surface everywhere on the surface, so the flux is positive.

9. True. In the sum defining the flux integral for $\vec{F}$, we have terms like $\vec{F} \cdot \Delta\vec{A} = (2\vec{G}) \cdot \Delta\vec{A} = 2(\vec{G} \cdot \Delta\vec{A})$. So each term in the sum approximating the flux of $\vec{F}$ is twice the corresponding term in the sum approximating the flux of $\vec{G}$, making the sum for $\vec{F}$ twice that of the sum for $\vec{G}$. Thus the flux of $\vec{F}$ is twice the flux of $\vec{G}$.

13. False. Both surfaces are oriented upward, so $\vec{A}(x,y)$ and $\vec{B}(x,y)$ both point upward. But they could point in different directions, since the graph of $z = -f(x,y)$ is the graph of $z = f(x,y)$ turned upside down.

CHAPTER TWENTY

Solutions for Section 20.1

Exercises

1. Two vector fields that have positive divergence everywhere are as follows:

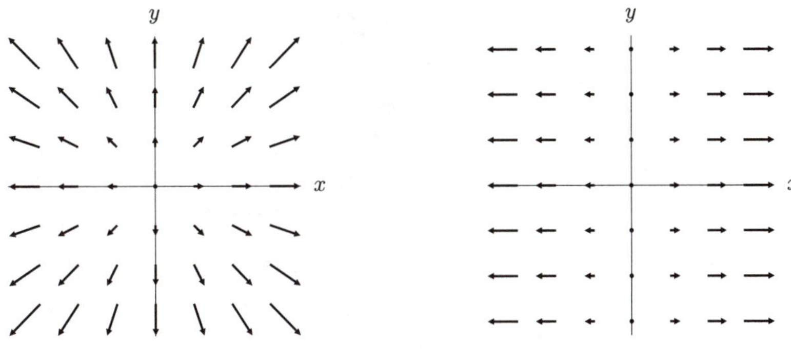

Figure 20.1

Figure 20.2

5. $\operatorname{div}\vec{F} = \frac{\partial}{\partial x}(-y) + \frac{\partial}{\partial y}(x) = 0$

9. In coordinates, we have

$$\vec{F}(x,y,z) = \frac{(x-x_0)}{\sqrt{(x-x_0)^2+(y-y_0)^2+(z-z_0)^2}}\vec{i} + \frac{(y-y_0)}{\sqrt{(x-x_0)^2+(y-y_0)^2+(z-z_0)^2}}\vec{j} + \frac{(z-z_0)}{\sqrt{(x-x_0)^2+(y-y_0)^2+(z-z_0)^2}}\vec{k}.$$

So if $(x,y,z) \neq (x_0,y_0,z_0)$, then

$$\begin{aligned}
\operatorname{div}\vec{F} &= \left(\frac{1}{\sqrt{(x-x_0)^2+(y-y_0)^2+(z-z_0)^2}} - \frac{(x-x_0)^2}{((x-x_0)^2+(y-y_0)^2+(z-z_0)^2)^{3/2}}\right) \\
&+ \left(\frac{1}{\sqrt{(x-x_0)^2+(y-y_0)^2+(z-z_0)^2}} - \frac{(y-y_0)^2}{((x-x_0)^2+(y-y_0)^2+(z-z_0)^2)^{3/2}}\right) \\
&+ \left(\frac{1}{\sqrt{(x-x_0)^2+(y-y_0)^2+(z-z_0)^2}} - \frac{(z-z_0)^2}{((x-x_0)^2+(y-y_0)^2+(z-z_0)^2)^{3/2}}\right) \\
&= \left(\frac{(x-x_0)^2+(y-y_0)^2+(z-z_0)^2}{((x-x_0)^2+(y-y_0)^2+(z-z_0)^2)^{3/2}} - \frac{(x-x_0)^2}{((x-x_0)^2+(y-y_0)^2+(z-z_0)^2)^{3/2}}\right) \\
&+ \left(\frac{(x-x_0)^2+(y-y_0)^2+(z-z_0)^2}{((x-x_0)^2+(y-y_0)^2+(z-z_0)^2)^{3/2}} - \frac{(y-y_0)^2}{((x-x_0)^2+(y-y_0)^2+(z-z_0)^2)^{3/2}}\right) \\
&+ \left(\frac{(x-x_0)^2+(y-y_0)^2+(z-z_0)^2}{((x-x_0)^2+(y-y_0)^2+(z-z_0)^2)^{3/2}} - \frac{(z-z_0)^2}{((x-x_0)^2+(y-y_0)^2+(z-z_0)^2)^{3/2}}\right) \\
&= \frac{3((x-x_0)^2+(y-y_0)^2+(z-z_0)^2) - ((x-x_0)^2+(y-y_0)^2+(z-z_0)^2)}{((x-x_0)^2+(y-y_0)^2+(z-z_0)^2)^{3/2}} \\
&= \frac{2}{\sqrt{(x-x_0)^2+(y-y_0)^2+(z-z_0)^2}} = \frac{2}{\|\vec{r}-\vec{r}_0\|}.
\end{aligned}$$

Problems

13. Since div $F(1,2,3)$ is the flux density out of a small region surrounding the point $(1,2,3)$, we have

$$\operatorname{div}\vec{F}\,(1,2,3) \approx \frac{\text{Flux out of small region around } (1,2,3)}{\text{Volume of region.}}$$

So

$$\begin{aligned}\text{Flux out of region} &\approx (\operatorname{div}\vec{F}\,(1,2,3)) \cdot \text{Volume of region}\\ &= 5 \cdot \frac{4}{3}\pi(0.01)^3\\ &= \frac{0.00002\pi}{3}.\end{aligned}$$

17.

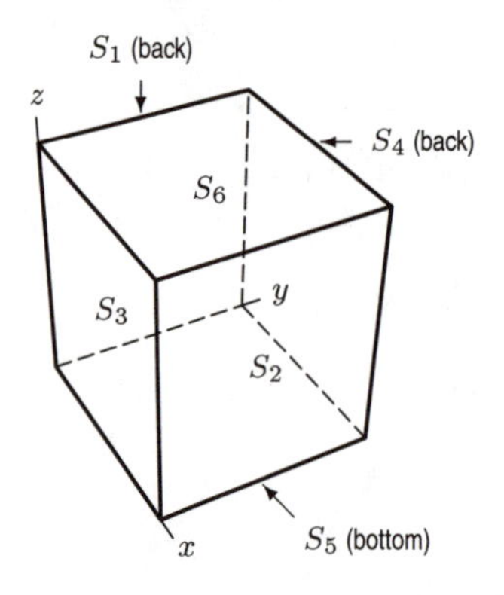

Figure 20.3

(a) The vector field is parallel to the x-axis and zero on the yz-plane. Thus the only contribution to the flux is from S_2. On S_2, $x = c$, the normal is outward. Since $\vec{F}$ is constant on S_2, the flux through face S_2 is

$$\begin{aligned}\int_{S_2} \vec{F} \cdot d\vec{A} &= \vec{F} \cdot \vec{A}_{S_2}\\ &= c\vec{i} \cdot c^2\vec{i}\\ &= c^3.\end{aligned}$$

Thus, total flux through box $= c^3$.

(b) Using the geometric definition of convergence

$$\begin{aligned}\operatorname{div}\vec{F} &= \lim_{c\to 0}\left(\frac{\text{Flux through box}}{\text{Volume of box}}\right)\\ &= \lim_{c\to 0}\left(\frac{c^3}{c^3}\right)\\ &= 1\end{aligned}$$

(c)

$$\frac{\partial}{\partial x}(x) + \frac{\partial}{\partial y}(0) + \frac{\partial}{\partial z}(0) = 1 + 0 + 0 = 1.$$

21. Let $\vec{a} = a_1\vec{i} + a_2\vec{j} + a_3\vec{k}$ with a_1, a_2, and a_3 constant. Then $f\vec{a} = f(x,y,z)(a_1\vec{i} + a_2\vec{j} + a_3\vec{k}) = f(x,y,z)a_1\vec{i} + f(x,y,z)a_2\vec{j} + f(x,y,z)a_3\vec{k} = fa_1\vec{i} + fa_2\vec{j} + fa_3\vec{k}$. So

$$\begin{aligned}
\operatorname{div}(f\vec{a}) &= \frac{\partial(fa_1)}{\partial x} + \frac{\partial(fa_2)}{\partial y} + \frac{\partial(fa_3)}{\partial z} \\
&= a_1\frac{\partial f}{\partial x} + a_2\frac{\partial f}{\partial y} + a_3\frac{\partial f}{\partial z} \quad \text{since } a_1, a_2, a_3 \text{ are constants} \\
&= (\frac{\partial f}{\partial x}\vec{i} + \frac{\partial f}{\partial y}\vec{j} + \frac{\partial f}{\partial z}\vec{k}) \cdot (a_1\vec{i} + a_2\vec{j} + a_3\vec{k}) \\
&= (\operatorname{grad} f) \cdot \vec{a}.
\end{aligned}$$

25. Calculate $r_x = (1/2)(x^2+y^2)^{-1/2}2x = x/r$ and $r_y = y/r$. We have

$$\begin{aligned}
\operatorname{div} \frac{h(r)}{r^2}(x\vec{i} + y\vec{j}) &= \frac{\partial}{\partial x}\frac{xh(r)}{r^2} + \frac{\partial}{\partial y}\frac{yh(r)}{r^2} \\
&= \frac{h(r)}{r^2} + x\frac{r^2h'(r)r_x - 2h(r)rr_x}{r^4} + \frac{h(r)}{r^2} + y\frac{r^2h'(r)r_y - 2h(r)rr_y}{r^4} \\
&= 2\frac{h(r)}{r^2} + \left(\frac{h'(r)}{r^2} - \frac{2h(r)}{r^3}\right)(xr_x + yr_y) \\
&= 2\frac{h(r)}{r^2} + \left(\frac{h'(r)}{r^2} - \frac{2h(r)}{r^3}\right)\left(\frac{x^2+y^2}{r}\right) \\
&= 2\frac{h(r)}{r^2} + \left(\frac{h'(r)}{r^2} - \frac{2h(r)}{r^3}\right)\left(\frac{r^2}{r}\right) \\
&= \frac{h'(r)}{r}.
\end{aligned}$$

29. **(a)** Translating the vector field into rectangular coordinates gives, if $(x,y,z) \neq (0,0,0)$

$$\vec{E}(x,y,z) = \frac{kx}{(x^2+y^2+z^2)^{3/2}}\vec{i} + \frac{ky}{(x^2+y^2+z^2)^{3/2}}\vec{j} + \frac{kz}{(x^2+y^2+z^2)^{3/2}}\vec{k}.$$

We now take the divergence of this to get

$$\begin{aligned}
\operatorname{div}\vec{E} &= k\left(-3\frac{x^2+y^2+z^2}{(x^2+y^2+z^2)^{5/2}} + \frac{3}{(x^2+y^2+z^2)^{3/2}}\right) \\
&= 0.
\end{aligned}$$

(b) Let S be the surface of a sphere centered at the origin. We have seen that for this field, the flux $\int \vec{E}\cdot d\vec{A}$ is the same for all such spheres, regardless of their radii. So let the constant c stand for $\int \vec{E}\cdot d\vec{A}$. Then

$$\operatorname{div}\vec{E}(0,0,0) = \lim_{\text{vol}\to 0}\frac{\int \vec{E}\cdot d\vec{A}}{\text{Volume inside } S} = \lim_{\text{vol}\to 0}\frac{c}{\text{Volume}}.$$

(c) For a point charge, the charge density is not defined. The charge density is 0 everywhere else.

Solutions for Section 20.2

Exercises

1. No, because the surface S is not a closed surface.

5. First directly, since the vector field is totally in the $\vec{j}$ direction, there is no flux through the ends. On the side of the cylinder, a normal vector at (x, y, z) is $x\vec{i} + y\vec{j}$. This is in fact a unit normal, since $x^2 + y^2 = 1$ (the cylinder has radius 1). Also, using $x = \cos\theta, y = \sin\theta$, in this case, the element of area dA equals $1 d\theta dz$. So

$$\begin{aligned}\text{Flux} &= \int \vec{F} \cdot d\vec{A} = \int_0^2 \int_0^{2\pi} (y\vec{j}) \cdot (x\vec{i} + y\vec{j})\, d\theta\, dz \\ &= \int_0^2 \int_0^{2\pi} y^2\, d\theta\, dz = \int_0^2 \int_0^{2\pi} \sin^2\theta\, d\theta\, dz = \int_0^2 \pi dz = 2\pi.\end{aligned}$$

Now we calculate the flux using the divergence theorem. The divergence of the field is given by the sum of the respective partials of the components, so the divergence is simply $\dfrac{\partial y}{\partial y} = 1$. Since the divergence is constant, we can simply calculate the volume of the cylinder and multiply by the divergence

$$\text{Flux} = 1\pi r^2 h = 2\pi$$

Problems

9. Apply the Divergence Theorem to the solid cone, whose interior we call W. The surface of W consists of S and D. Thus

$$\int_S \vec{F} \cdot d\vec{A} + \int_D \vec{F} \cdot d\vec{A} = \int_W \operatorname{div} \vec{F}\, dV.$$

But $\operatorname{div} \vec{F} = 0$ everywhere, since $\vec{F}$ is constant. Thus

$$\int_D \vec{F} \cdot d\vec{A} = -\int_S \vec{F} \cdot d\vec{A} = -3.22.$$

13. **(a)** True. Flux that goes in one face goes out the other, since the vector field is constant and the surface is closed.
(b) True. The flux out of S_1 along the face shared with S_2 cancels with the flux out of S_2 over the same face. (The normals are in opposite directions.) The other five faces of S_1 and the other five faces of S_2 are each faces of S.

17. Any closed surface, S, oriented inward, will work. Then,

$$\int_{S(\text{inward})} \vec{F} \cdot d\vec{A} = -\int_{S(\text{outward})} \vec{F} \cdot d\vec{A},$$

so, by the Divergence Theorem, with W representing the region inside S,

$$\int_{S(\text{inward})} \vec{F} \cdot d\vec{A} = -\int_W \operatorname{div} \vec{F}\, dV = -\int_W (x^2 + y^2 + 3) dV.$$

The integral on the right is positive because the integrand is positive everywhere. Therefore the flux through S oriented inward is negative.

21. **(a)** We cannot use the Divergence Theorem to calculate the flux through the sphere of radius 2 because $\vec{G}$ is not defined throughout the interior of the sphere. We calculate the flux directly. Since $\vec{G}$ is parallel to the area vector at the surface of the sphere, and since $||\vec{G}|| = 4 \cdot 2^2 \cdot 2 = 32$ on the surface, we have

$$\text{Flux} = \int_S \vec{G} \cdot d\vec{A} = 32 \cdot \text{Area of surface} = 32 \cdot 4\pi 2^2 = 512\pi.$$

(b) The fact that the vectors of $\vec{G}$ get longer as we go away from the origin suggests that $\operatorname{div} \vec{G} > 0$. This is confirmed by calculating the divergence using the formula

$$\vec{G} = 4r^2 \vec{r} = 4(x^2 + y^2 + z^2)(x\vec{i} + y\vec{j} + z\vec{k}).$$

Since $\dfrac{\partial G_1}{\partial x} = 4(3x^2 + y^2 + z^2)$, and so on, $\operatorname{div} \vec{G}$ is positive everywhere outside the unit sphere.

The sphere is completely contained within the box. Apply the Divergence Theorem to the region, W, between the sphere and the box. This region has surface area the sphere (oriented inward) and the box (oriented outward). The Divergence Theorem gives

$$\int_{\text{Box (outward)}} \vec{F} \cdot d\vec{A} + \int_{\text{Sphere (inward)}} \vec{F} \cdot d\vec{A} = \int_W \operatorname{div} \vec{G}\, dV > 0.$$

So

$$\int_{\text{Box (outward)}} \vec{F} \cdot d\vec{A} - \int_{\text{Sphere (outward)}} \vec{F} \cdot d\vec{A} = \int_W \operatorname{div} \vec{G}\, dV > 0.$$

Thus,

$$\int_{\text{Box (outward)}} \vec{F} \cdot d\vec{A} = \int_{\text{Sphere (outward)}} \vec{F} \cdot d\vec{A} + \int_W \operatorname{div} \vec{G}\, dV.$$

So the flux through the box is larger than the flux through the sphere.

25. **(a)** Taking partial derivatives of $\vec{E}$ gives

$$\begin{aligned}\frac{\partial E_1}{\partial x} &= \frac{\partial}{\partial x}[qx(x^2+y^2+z^2)^{-3/2}] = q[(x^2+y^2+z^2)^{-3/2} + x(-3/2)(2x)(x^2+y^2+z^2)^{-5/2}] \\ &= q(y^2+z^2-2x^2)(x^2+y^2+z^2)^{-5/2}.\end{aligned}$$

Similarly,

$$\begin{aligned}\frac{\partial E_2}{\partial x} &= q(x^2+z^2-2y^2)(x^2+y^2+z^2)^{-5/2} \\ \frac{\partial E_3}{\partial x} &= q(x^2+y^2-2z^2)(x^2+y^2+z^2)^{-5/2}.\end{aligned}$$

Summing, we obtain $\operatorname{div} \vec{E} = 0$.

(b) Since on the surface of the sphere, the vector field $\vec{E}$ and the area vector $\Delta\vec{A}$ are parallel,

$$\vec{E} \cdot \Delta\vec{A} = \|\vec{E}\|\|\Delta\vec{A}\|.$$

Now, on the surface of a sphere of radius a,

$$\|\vec{E}\| = \frac{q\|\vec{r}\|}{\|\vec{r}\|^3} = \frac{q}{a^2}.$$

Thus,

$$\int_{S_a} \vec{E} \cdot d\vec{A} = \int \frac{q}{a^2}\|d\vec{A}\| = \frac{q}{a^2} \cdot \text{Surface area of sphere} = \frac{q}{a^2} \cdot 4\pi a^2 = 4\pi q.$$

(c) It is not possible to apply the Divergence Theorem in part (b) since $\vec{E}$ is not defined at the origin (which lies inside the region of space bounded by S_a), and the Divergence Theorem requires that the vector field be defined everywhere inside S.

(d) Let R be the solid region lying between a small sphere S_a, centered at the origin, and the surface S. Applying the Divergence Theorem and the result of part (a), we get:

$$0 = \int_R \operatorname{div} \vec{E}\, dV = \int_{S_a} \vec{E} \cdot d\vec{A} + \int_S \vec{E} \cdot d\vec{A},$$

where S is oriented with the outward normal vector, and S_a with the inward normal vector (since this is "outward" with respect to the region R). Since

$$\int_{S_a,\text{ inward}} \vec{E} \cdot d\vec{A} = -\int_{S_a,\text{ outward}} \vec{E} \cdot d\vec{A},$$

the result of part (b) yields

$$\int_S \vec{E} \cdot d\vec{A} = 4\pi q.$$

[Note: It is legitimate to apply the Divergence Theorem to the region R since the vector field $\vec{E}$ is defined everywhere in R.]

Solutions for Section 20.3

Exercises

1. Using the definition in Cartesian coordinates, we have

$$
\begin{aligned}
\operatorname{curl}\vec F &= \begin{vmatrix} \vec i & \vec j & \vec k \\ \frac{\partial}{\partial x} & \frac{\partial}{\partial y} & \frac{\partial}{\partial z} \\ x^2-y^2 & 2xy & 0 \end{vmatrix} \\
&= \left(\frac{\partial}{\partial y}(0) - \frac{\partial}{\partial z}(2xy)\right)\vec i + \left(-\frac{\partial}{\partial x}(0) + \frac{\partial}{\partial z}(x^2-y^2)\right)\vec j + \left(\frac{\partial}{\partial x}(2xy) - \frac{\partial}{\partial y}(x^2-y^2)\right)\vec k \\
&= 4y\vec k\,.
\end{aligned}
$$

5. Using the definition of Cartesian coordinates,

$$
\begin{aligned}
\operatorname{curl}\vec F &= \begin{vmatrix} \vec i & \vec j & \vec k \\ \frac{\partial}{\partial x} & \frac{\partial}{\partial y} & \frac{\partial}{\partial z} \\ (-x+y) & (y+z) & (-z+x) \end{vmatrix} \\
&= \left(\frac{\partial}{\partial y}(-z+x) - \frac{\partial}{\partial z}(y+z)\right)\vec i + \left(-\frac{\partial}{\partial x}(-z+x) + \frac{\partial}{\partial z}(-x+y)\right)\vec j \\
&\quad + \left(\frac{\partial}{\partial x}(y+z) - \frac{\partial}{\partial y}(-x+y)\right)\vec k \\
&= -\vec i - \vec j - \vec k\,.
\end{aligned}
$$

9. This vector field shows no rotation, and the circulation around any closed curve appears to be zero, so we suspect a zero curl here.

Problems

13. The conjecture is that when the first component of $\vec F$ depends only on x, the second component depends only on y, and the third component depends only on z, that is, if

$$\vec F = F_1(x)\vec i + F_2(y)\vec j + F_3(z)\vec k$$

then

$$\operatorname{curl}\vec F = \vec 0$$

The reason for this is that if $\vec F = F_1(x)\vec i + F_2(y)\vec j + F_3(z)\vec k$, then

$$
\begin{aligned}
\operatorname{curl}\vec F &= \begin{vmatrix} \vec i & \vec j & \vec k \\ \frac{\partial}{\partial x} & \frac{\partial}{\partial y} & \frac{\partial}{\partial z} \\ F_1(x) & F_2(y) & F_3(z) \end{vmatrix} \\
&= \left(\frac{\partial}{\partial y}F_3(z) - \frac{\partial}{\partial z}F_2(y)\right)\vec i + \left(-\frac{\partial}{\partial x}F_3(z) + \frac{\partial}{\partial z}F_1(x)\right)\vec j + \left(\frac{\partial}{\partial x}F_2(y) - \frac{\partial}{\partial y}F_1(x)\right)\vec k \\
&= \vec 0\,.
\end{aligned}
$$

17. Investigate the velocity vector field of the atmosphere near the fire. If the curl of this vector field is non-zero, there is circulatory motion. Consequently, if the magnitude of the curl of this vector field is large near the fire, a fire storm has probably developed.

21. The vector curl $\vec{F}$ has its component in the x-direction given by

$$(\text{curl}\,\vec{F})_x \approx \frac{\text{Circulation around small circle around } x\text{-axis}}{\text{Area inside circle}}$$

$$= \frac{\text{Circulation around } C_2}{\text{Area inside } C_2} = \frac{0.5\pi}{\pi(0.1)^2} = 50.$$

Similar reasoning leads to

$$(\text{curl}\,\vec{F})_y \approx \frac{\text{Circulation around } C_3}{\text{Area inside } C_3} = \frac{3\pi}{\pi(0.1)^2} = 300,$$

$$(\text{curl}\,\vec{F})_z \approx \frac{\text{Circulation around } C_1}{\text{Area inside } C_1} = \frac{0.02\pi}{\pi(0.1)^2} = 2.$$

Thus,

$$\text{curl}\,\vec{F} \approx 50\vec{i} + 300\vec{j} + 2\vec{k}.$$

25. Let $\vec{c} = c_1\vec{i} + c_2\vec{j} + c_3\vec{k}$ and $\vec{F} = F_1\vec{i} + F_2\vec{j} + F_3\vec{k}$. We then show the desired result as follows:

$$\begin{aligned}
\text{div}(\vec{F} \times \vec{c}) &= \text{div}((F_1\vec{i} + F_2\vec{j} + F_3\vec{k}) \times (c_1\vec{i} + c_2\vec{j} + c_3\vec{k}))\\
&= \text{div}((F_2c_3 - F_3c_2)\vec{i} + (F_3c_1 - F_1c_3)\vec{j} + (F_1c_2 - F_2c_1)\vec{k})\\
&= \frac{\partial}{\partial x}(F_2c_3 - F_3c_2) + \frac{\partial}{\partial y}(F_3c_1 - F_1c_3) + \frac{\partial}{\partial z}(F_1c_2 - F_2c_1)\\
&= c_3\frac{\partial F_2}{\partial x} - c_2\frac{\partial F_3}{\partial x} + c_1\frac{\partial F_3}{\partial y} - c_3\frac{\partial F_1}{\partial y} + c_2\frac{\partial F_1}{\partial z} - c_1\frac{\partial F_2}{\partial z}\\
&= c_1(\frac{\partial F_3}{\partial y} - \frac{\partial F_2}{\partial z}) + c_2(\frac{\partial F_1}{\partial z} - \frac{\partial F_3}{\partial x}) + c_3(\frac{\partial F_2}{\partial x} - \frac{\partial F_1}{\partial y})\\
&= (c_1\vec{i} + c_2\vec{j} + c_3\vec{k}) \cdot ((\frac{\partial F_3}{\partial y} - \frac{\partial F_2}{\partial z})\vec{i} + (\frac{\partial F_1}{\partial z} - \frac{\partial F_3}{\partial x})\vec{j} + (\frac{\partial F_2}{\partial x} - \frac{\partial F_1}{\partial y})\vec{k})\\
&= \vec{c} \cdot \text{curl}\,\vec{F}.
\end{aligned}$$

29. (a) Figure 20.4 shows a cross-section of the vector field in xy-plane with $\omega = 1$, so $\vec{v} = -y\vec{i} + x\vec{j}$.

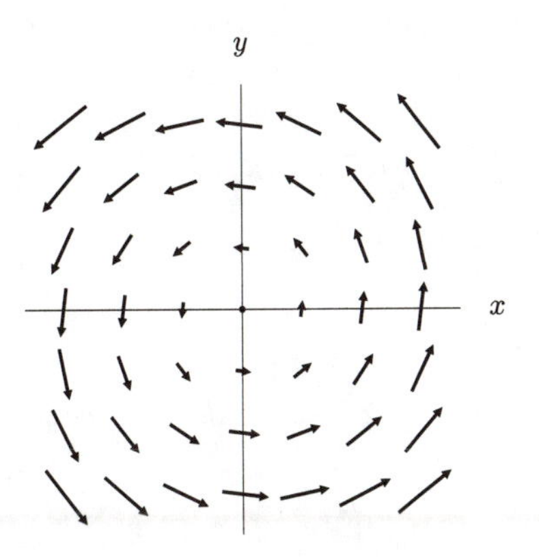

Figure 20.4: $\vec{v} = -y\vec{i} + x\vec{j}$

Figure 20.5 shows a cross-section of vector field in xy-plane with $\omega = -1$, so $\vec{v} = y\vec{i} - x\vec{j}$.

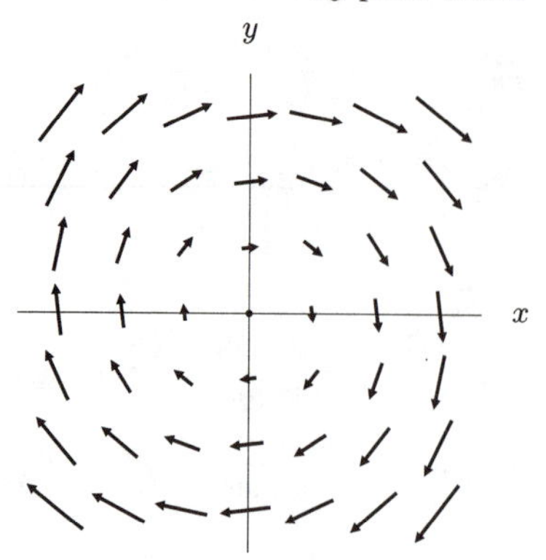

Figure 20.5: $\vec{v} = y\vec{i} - x\vec{j}$

(b) The distance from the center of the vortex is given by $r = \sqrt{x^2 + y^2}$. The velocity of the vortex at any point is $-\omega y\vec{i} + \omega x\vec{j}$, and the speed of the vortex at any point is the magnitude of the velocity, or $s = ||\vec{v}|| = \sqrt{(-\omega y)^2 + (\omega x)^2} = |\omega| \sqrt{x^2 + y^2} = |\omega| r$.

(c) The divergence of the velocity field is given by:

$$\operatorname{div} \vec{v} = \frac{\partial(-\omega y)}{\partial x} + \frac{\partial(\omega x)}{\partial y} = 0$$

The curl of the field is:

$$\operatorname{curl} \vec{v} = \operatorname{curl}(-\omega y\vec{i} + \omega x\vec{j}) = \left(\frac{\partial}{\partial x}(\omega x) - \frac{\partial}{\partial y}(-\omega y)\right)\vec{k} = 2\omega\vec{k}$$

(d) We know that $\vec{v}$ has constant magnitude $|\omega|R$ everywhere on the circle and is everywhere tangential to the circle. In addition, if $\omega > 0$, the vector field rotates counterclockwise; if $\omega < 0$, the vector field rotates clockwise. Thus if $\omega > 0$, $\vec{v}$ and $\Delta\vec{r}$ are parallel and in the same direction, so

$$\int_C \vec{v} \cdot d\vec{r} = |\vec{v}| \cdot (\text{Length of } C) = \omega R \cdot 2\pi R = 2\pi\omega R^2$$

If $\omega < 0$, then $|\omega| = -\omega$ and $\vec{v}$ and $\Delta\vec{r}$ are in opposite directions, so

$$\int_C \vec{v} \cdot d\vec{r} = -|\vec{v}| \cdot (\text{Length of } C) = -|\omega| R \cdot (2\pi R) = 2\pi\omega R^2.$$

33. Let $\vec{v} = a\vec{i} + b\vec{j} + c\vec{k}$ and try

$$\vec{F} = \vec{v} \times \vec{r} = (a\vec{i} + b\vec{j} + c\vec{k}) \times (x\vec{i} + y\vec{j} + z\vec{k}) = (bz - cy)\vec{i} + (cx - az)\vec{j} + (ay - bx)\vec{k}.$$

Then

$$\operatorname{curl} \vec{F} = \begin{vmatrix} \vec{i} & \vec{j} & \vec{k} \\ \frac{\partial}{\partial x} & \frac{\partial}{\partial y} & \frac{\partial}{\partial z} \\ bz - cy & cx - az & ay - bx \end{vmatrix} = 2a\vec{i} + 2b\vec{j} + 2c\vec{k}.$$

Taking $a = 1$, $b = -\frac{3}{2}$, $c = 2$ gives $\operatorname{curl}\vec{F} = 2\vec{i} - 3\vec{j} + 4\vec{k}$, so the desired vector field is $\vec{F} = (-\frac{3}{2}z - 2y)\vec{i} + (2x - z)\vec{j} + (y + \frac{3}{2}x)\vec{k}$.

Solutions for Section 20.4

Exercises

1. No, because the curve C over which the integral is taken is not a closed curve, and so it is not the boundary of a surface.

5. The graph of $\vec{F} = \vec{r}/||\vec{r}||^3$ consists of vectors pointing radially outward. There is no swirl, so $\operatorname{curl}\vec{F} = \vec{0}$. From Stokes' Theorem,

$$\int_C \vec{F} \cdot d\vec{r} = \int_S \operatorname{curl}\vec{F} \cdot d\vec{A} = \int_S \vec{0} \cdot d\vec{A} = 0$$

Problems

9. **(a)** It appears that div $\vec{F} < 0$, and div $\vec{G} < 0$; div $\vec{G}$ is larger in magnitude (more negative) if the scales are the same.
(b) curl $\vec{F}$ and curl $\vec{G}$ both appear to be zero at the origin (and elsewhere).
(c) Yes, the cylinder with axis along the z-axis will have negative flux through it (ends parallel to xy-plane).
(d) Same as part(c).
(e) No, you cannot draw a closed curve around the origin such that $\vec{F}$ has a non-zero circulation around it because curl is zero. By Stokes' theorem, circulation equals the integral of the curl over the surface bounded by the curve.
(f) Same as part(e)

13. **(a)** We calculate

$$\text{curl}\,\vec{F} = \begin{vmatrix} \vec{i} & \vec{j} & \vec{k} \\ \frac{\partial}{\partial x} & \frac{\partial}{\partial y} & \frac{\partial}{\partial z} \\ y-z & x+z & xy \end{vmatrix} = (x-1)\vec{i} - (y+1)\vec{j} + (1-1)\vec{k} = (x-1)\vec{i} - (y+1)\vec{j}.$$

Since the circle is in the xy-plane and curl $\vec{F}$ has no $\vec{k}$ component, the line integral is zero.
(b) Even though the line integral around this closed curve is 0, we do not know that the line integral around every closed curve is zero, so we cannot conclude that $\vec{F}$ is path-independent (conservative). Since curl $\vec{F} \neq \vec{0}$, we know that $\vec{F}$ is not path-independent (conservative).

17. **(a)** Computing curl $\vec{G}$, we get

$$\begin{aligned}\text{curl}\,\vec{G} &= \left(\frac{\partial(xz)}{\partial y} - \frac{\partial(3x)}{\partial z}\right)\vec{i} + \left(\frac{\partial(4yz^2)}{\partial z} - \frac{\partial(xz)}{\partial x}\right)\vec{j} + \left(\frac{\partial(3x)}{\partial x} - \frac{\partial(4yz^2)}{\partial y}\right)\vec{k} \\ &= 0\vec{i} + (8yz - z)\vec{j} + (3 - 4z^2)\vec{k} \\ &= \vec{F}\end{aligned}$$

(b) Because $\vec{F}$ is a curl field, the flux of $\vec{F}$ through S is the same as its flux through any surface with the same boundary as S. Let us replace S by the circular disk S_1 of radius 5 in the xy-plane. On S_1, $z = 0$, and so the vector field reduces to $\vec{F} = 3\vec{k}$. Since the vector $3\vec{k}$ is normal to S_1 and in the direction of the orientation of S_1, we have

$$\int_S \vec{F} \cdot d\vec{A} = \int_{S_1} \vec{F} \cdot d\vec{A} = \int_{S_1} 3\vec{k} \cdot d\vec{A} = \|3\vec{k}\|(\text{area of } S_1) = 75\pi.$$

21. Notice that $\vec{F}$, div $\vec{F}$, and curl $\vec{F}$ are defined and continuous everywhere.

(a) By the Divergence Theorem, the fact that the flux through any closed surface is 0 tells us that everywhere

$$\text{div}\,\vec{F} = 0.$$

Since div $\vec{F} = a$, we know $a = 0$. We do not have any information about b, c, m.
(b) Since the circulation around any closed curve is 0, by Stokes' Theorem, we have

$$\text{curl}\,\vec{F} = \vec{0}.$$

Now

$$\text{curl}\,\vec{F} = \begin{vmatrix} \vec{i} & \vec{j} & \vec{k} \\ \frac{\partial}{\partial x} & \frac{\partial}{\partial y} & \frac{\partial}{\partial z} \\ ax+by+5z & x+cz & 3y+mx \end{vmatrix} = (3-c)\vec{i} - (m-5)\vec{j} + (1-b)\vec{k},$$

so curl $\vec{F} = 0$ means $c = 3, m = 5$ and $b = 1$. We do not have any information about a.

Solutions for Section 20.5

Exercises

1. Since curl $\vec{F} = \vec{0}$ and $\vec{F}$ is defined everywhere, we know by the curl test that $\vec{F}$ is a gradient field. In fact, $\vec{F} = \text{grad} f$, where $f(x, y, z) = xyz + yz^2$, so f is a potential function for $\vec{F}$.

Problems

5. We must show $\text{curl}\,\vec{A} = \vec{B}$.

$$\begin{aligned}\text{curl}\,\vec{A} &= \frac{\partial}{\partial y}\left(\frac{-I}{c}\ln(x^2+y^2)\right)\vec{i} - \frac{\partial}{\partial x}\left(\frac{-I}{c}\ln(x^2+y^2)\right)\vec{j} \\ &= \frac{-I}{c}\left(\frac{2y}{x^2+y^2}\right)\vec{i} + \frac{I}{c}\left(\frac{2x}{(x^2+y^2)}\right)\vec{j} \\ &= \frac{2I}{c}\left(\frac{-y\vec{i}+x\vec{j}}{x^2+y^2}\right) \\ &= \vec{B}.\end{aligned}$$

9. (a) Yes. To show this, we use a version of the product rule for curl (Problem 26 on page 327):

$$\text{curl}(\phi\vec{F}) = \phi\,\text{curl}\,\vec{F} + (\text{grad}\,\phi)\times\vec{F},$$

where ϕ is a scalar function and $\vec{F}$ is a vector field. So

$$\begin{aligned}\text{curl}\left(q\frac{\vec{r}}{\|\vec{r}\|^3}\right) = \text{curl}\left(\frac{q}{\|\vec{r}\|^3}\vec{r}\right) &= \frac{q}{\|\vec{r}\|^3}\,\text{curl}\,\vec{r} + \text{grad}\left(\frac{q}{\|\vec{r}\|^3}\right)\times\vec{r} \\ &= \vec{0} + q\,\text{grad}\left(\frac{1}{\|\vec{r}\|^3}\right)\times\vec{r}\end{aligned}$$

Since the level surfaces of $1/\|\vec{r}\|^3$ are spheres centered at the origin, $\text{grad}(1/\|\vec{r}\|^3)$ is parallel to $\vec{r}$, so $\text{grad}(1/\|\vec{r}\|^3)\times\vec{r} = \vec{0}$. Thus, $\text{curl}\,\vec{E} = \vec{0}$.

(b) Yes. The domain of $\vec{E}$ is 3-space minus $(0,0,0)$. Any closed curve in this region is the boundary of a surface contained entirely in the region. (If the first surface you pick happens to contain $(0,0,0)$, change its shape slightly to avoid it.)

(c) Yes. Since $\vec{E}$ satisfies both conditions of the curl test, it must be a gradient field. In fact,

$$\vec{E} = \text{grad}\left(-q\frac{1}{\|\vec{r}\|}\right).$$

13. (a) Since $\text{curl}\,\text{grad}\,\psi = 0$ for any function ψ, $\text{curl}(\vec{A}+\text{grad}\,\psi) = \text{curl}\,\vec{A} + \text{curl}\,\text{grad}\,\psi = \text{curl}\,\vec{A} = \vec{B}$.

(b) We have

$$\text{div}(\vec{A}+\text{grad}\,\psi) = \text{div}\,\vec{A} + \text{div}\,\text{grad}\,\psi = \text{div}\,\vec{A} + \nabla^2\psi.$$

Thus ψ should be chosen to satisfy the partial differential equation

$$\nabla^2\psi = -\,\text{div}\,\vec{A}.$$

Solutions for Chapter 20 Review

Exercises

1. (a) We have

$$\text{curl}\,\vec{F} = \begin{vmatrix}\vec{i} & \vec{j} & \vec{k} \\ \frac{\partial}{\partial x} & \frac{\partial}{\partial y} & \frac{\partial}{\partial z} \\ \cos x & e^y & x-y-z\end{vmatrix} = -\vec{i} - \vec{j}.$$

(b) If S is the disk on the plane within the circle C, Stokes' Theorem gives

$$\int_C \vec{F} \cdot d\vec{r} = \int_S \operatorname{curl} \vec{F} \cdot d\vec{A}.$$

For Stokes' Theorem, the disk is oriented upward. Since the unit normal to the plane is $(\vec{i} + \vec{j} + \vec{k})/\sqrt{3}$ and the disk has radius 3, the area vector of the disk is

$$\vec{A} = \frac{\vec{i} + \vec{j} + \vec{k}}{\sqrt{3}} \pi(3^2) = 3\sqrt{3}\pi(\vec{i} + \vec{j} + \vec{k}).$$

Thus, using curl $\vec{F} = -\vec{i} - \vec{j}$, we have

$$\int_C \vec{F} \cdot d\vec{r} = (-\vec{i} - \vec{j}) \cdot 3\sqrt{3}\pi(\vec{i} + \vec{j} + \vec{k}) = -6\sqrt{3}\pi.$$

5. (a) We will compute separately the flux of the vector field $\vec{F} = x^3\vec{i} + 2y\vec{j} + 3\vec{k}$ through each of the six faces of the cube.

The face S_I where $x = 1$, which has normal vector $\vec{i}$. Only the $\vec{i}$ component $x^3\vec{i} = \vec{i}$ of $\vec{F}$ has flux through S_I.

$$\int_{S_I} \vec{F} \cdot d\vec{A} = \int_{S_I} \vec{i} \cdot d\vec{A} = \|\vec{i}\|(\text{area of } S_I) = 4.$$

The face S_{II} where $x = -1$, which has normal vector $-\vec{i}$. Only the $\vec{i}$ component $x^3\vec{i} = -\vec{i}$ of $\vec{F}$ has flux through S_{II}.

$$\int_{S_{II}} \vec{F} \cdot d\vec{A} = \int_{S_{II}} -\vec{i} \cdot d\vec{A} = \|-\vec{i}\|(\text{area of } S_{II}) = 4.$$

The face S_{III} where $y = 1$, which has normal vector $\vec{j}$. Only the $\vec{j}$ component $2y\vec{j} = 2\vec{j}$ of $\vec{F}$ has flux through S_{III}.

$$\int_{S_{III}} \vec{F} \cdot d\vec{A} = \int_{S_{III}} 2\vec{j} \cdot d\vec{A} = \|2\vec{j}\|(\text{area of } S_{III}) = 8.$$

The face S_{IV} where $y = -1$, which has normal vector $-\vec{j}$. Only the $\vec{j}$ component $2y\vec{j} = -2\vec{j}$ of $\vec{F}$ has flux through S_{IV}.

$$\int_{S_{IV}} \vec{F} \cdot d\vec{A} = \int_{S_{IV}} -2\vec{j} \cdot d\vec{A} = \|-2\vec{j}\|(\text{area of } S_{IV}) = 8.$$

The face S_V where $z = 1$, which has normal vector $\vec{k}$. Only the $\vec{k}$ component $3\vec{k}$ of $\vec{F}$ has flux through S_V.

$$\int_{S_V} \vec{F} \cdot d\vec{A} = \int_{S_V} 3\vec{k} \cdot d\vec{A} = \|3\vec{k}\|(\text{area of } S_V) = 12.$$

The face S_{VI} where $z = -1$, which has normal vector $-\vec{k}$. Only the $\vec{k}$ component $3\vec{k}$ of $\vec{F}$ has flux through S_{VI}.

$$\int_{S_{VI}} \vec{F} \cdot d\vec{A} = \int_{S_{VI}} 3\vec{k} \cdot d\vec{A} = -\|3\vec{k}\|(\text{area of } S_{VI}) = -12.$$

$$(\text{Total flux through } S) = 4 + 4 + 8 + 8 + 12 - 12 = 24.$$

(b) Since S is a closed surface the Divergence Theorem applies. Since $\operatorname{div}\vec{F} = 3x^2 + 2$,

$$\int_S \vec{F} \cdot d\vec{A} = \int_{x=-1}^{1} \int_{y=-1}^{1} \int_{z=-1}^{1} (3x^2 + 2)\,dz\,dy\,dx = 24.$$

9. C_2, C_3, C_4, C_6, since line integrals around C_1 and C_5 are clearly nonzero. You can see directly that $\int_{C_2} \vec{F} \cdot d\vec{r}$ and $\int_{C_6} \vec{F} \cdot d\vec{r}$ are zero, because C_2 and C_6 are perpendicular to their fields at every point.

13. We have

$$\operatorname{div} \vec{F} = \frac{\partial}{\partial x}(e^{y+z}) + \frac{\partial}{\partial y}(\sin(x+z)) + \frac{\partial}{\partial z}(x^2+y^2) = 0$$

$$\operatorname{curl} \vec{F} = \begin{vmatrix} \vec{i} & \vec{j} & \vec{k} \\ \frac{\partial}{\partial x} & \frac{\partial}{\partial y} & \frac{\partial}{\partial z} \\ e^{y+z} & \sin(x+z) & x^2+y^2 \end{vmatrix} = (2y - \cos(x+z))\vec{i} - \left(2x - e^{y+z}\right)\vec{j} + \left(\cos(x+z) - e^{y+z}\right)\vec{k}.$$

So $\vec{F}$ is solenoidal, but $\vec{F}$ is not irrotational.

17. Since $\operatorname{div}\vec{F} = 3x^2 + 3y^2$, using cylindrical coordinates to calculate the triple integral gives

$$\int_S \vec{F} \cdot d\vec{A} = \int_{\substack{\text{Interior}\\ \text{of cylinder}}} (3x^2+3y^2)\,dV = 3\int_0^{2\pi}\int_0^5\int_0^2 r^2 \cdot r\,dr\,dz\,d\theta = 3 \cdot 2\pi \cdot 5 \frac{r^4}{4}\bigg|_0^2 = 120\pi.$$

21. If C is the rectangular path around the rectangle, traversed counterclockwise when viewed from above, Stokes' Theorem gives

$$\int_S \operatorname{curl} \vec{F} \cdot d\vec{A} = \int_C \vec{F} \cdot d\vec{r}.$$

The $\vec{k}$ component of $\vec{F}$ does not contribute to the line integral, and the $\vec{j}$ component contributes to the line integral only along the segments of the curve parallel to the y-axis. Thus, if we break the line integral into four parts

$$\int_S \operatorname{curl} \vec{F} \cdot d\vec{A} = \int_{(0,0)}^{(3,0)} \vec{F} \cdot d\vec{r} + \int_{(3,0)}^{(3,2)} \vec{F} \cdot d\vec{r} + \int_{(3,2)}^{(0,2)} \vec{F} \cdot d\vec{r} + \int_{(0,2)}^{(0,0)} \vec{F} \cdot d\vec{r},$$

we see that the first and third integrals are zero, and we can replace $\vec{F}$ by its $\vec{j}$ component in the other two

$$\int_S \operatorname{curl} \vec{F} \cdot d\vec{A} = \int_{(3,0)}^{(3,2)} (x+7)\vec{j} \cdot d\vec{r} + \int_{(0,2)}^{(0,0)} (x+7)\vec{j} \cdot d\vec{r}.$$

Now $x = 3$ in the first integral and $x = 0$ in the second integral and the variable of integration is y in both, so

$$\int_S \operatorname{curl} \vec{F} \cdot d\vec{A} = \int_0^2 10\,dy + \int_2^0 7\,dy = 20 - 14 = 6.$$

Problems

25. We use Stokes' Theorem. Since

$$\operatorname{curl} \vec{F} = \begin{vmatrix} \vec{i} & \vec{j} & \vec{k} \\ \frac{\partial}{\partial x} & \frac{\partial}{\partial y} & \frac{\partial}{\partial z} \\ x+y & y+2z & z+3x \end{vmatrix} = -2\vec{i} - 3\vec{j} - \vec{k},$$

if S is the interior of the square, then

$$\int_C \vec{F} \cdot d\vec{r} = \int_S \operatorname{curl} \vec{F} \cdot d\vec{A} = \int_S (-2\vec{i} - 3\vec{j} - \vec{k}) \cdot d\vec{A}$$

Since the area vector of S is $49\vec{j}$, we have

$$\int_C \vec{F} \cdot d\vec{r} = \int_S (-2\vec{i} - 3\vec{j} - \vec{k}) \cdot d\vec{A} = -3\vec{j} \cdot 49\vec{j} = -147.$$

29. The flux of $\vec{E}$ through a small sphere of radius R around the point marked P is negative, because all the arrows are pointing into the sphere. The divergence at P is

$$\operatorname{div} \vec{E}(P) = \lim_{\text{vol}\to 0}\left(\frac{\int_S \vec{E} \cdot d\vec{A}}{\text{Volume of sphere}}\right) = \lim_{R\to 0}\left(\frac{\text{Negative number}}{\frac{4}{3}\pi R^3}\right) \le 0.$$

By a similar argument, the divergence at Q must be positive or zero.

33. **(a)** Since $\vec{F}$ is radial, it is everywhere parallel to the area vector, $\Delta\vec{A}$. Also, $\|\vec{F}\| = 1$ on the surface of the sphere $x^2+y^2+z^2=1$, so

$$\begin{aligned}\text{Flux through the sphere} &= \int_S \vec{F}\cdot d\vec{A} = \lim_{\|\Delta\vec{A}\|\to 0}\sum \vec{F}\cdot\Delta\vec{A}\\ &= \lim_{\|\Delta\vec{A}\|\to 0}\sum \|\vec{F}\|\|\Delta\vec{A}\| = \lim_{\|\Delta\vec{A}\|\to 0}\sum\|\Delta\vec{A}\|\\ &= \text{Surface area of sphere} = 4\pi\cdot 1^2 = 4\pi.\end{aligned}$$

(b) In Cartesian coordinates,

$$\vec{F}(x,y,z) = \frac{x}{(x^2+y^2+z^2)^{3/2}}\vec{i} + \frac{y}{(x^2+y^2+z^2)^{3/2}}\vec{j} + \frac{z}{(x^2+y^2+z^2)^{3/2}}\vec{k}.$$

So,

$$\begin{aligned}\operatorname{div}\vec{F}(x,y,z) &= \left(\frac{1}{(x^2+y^2+z^2)^{3/2}} - \frac{3x^2}{(x^2+y^2+z^2)^{5/2}}\right)\\ &+\left(\frac{1}{(x^2+y^2+z^2)^{3/2}} - \frac{3y^2}{(x^2+y^2+z^2)^{5/2}}\right)\\ &+\left(\frac{1}{(x^2+y^2+z^2)^{3/2}} - \frac{3z^2}{(x^2+y^2+z^2)^{5/2}}\right)\\ &= \left(\frac{x^2+y^2+z^2}{(x^2+y^2+z^2)^{5/2}} - \frac{3x^2}{(x^2+y^2+z^2)^{5/2}}\right)\\ &+\left(\frac{x^2+y^2+z^2}{(x^2+y^2+z^2)^{5/2}} - \frac{3y^2}{(x^2+y^2+z^2)^{5/2}}\right)\\ &+\left(\frac{x^2+y^2+z^2}{(x^2+y^2+z^2)^{5/2}} - \frac{3z^2}{(x^2+y^2+z^2)^{5/2}}\right)\\ &= \frac{3(x^2+y^2+z^2)-3(x^2+y^2+z^2)}{(x^2+y^2+z^2)^{5/2}}\\ &= 0.\end{aligned}$$

(c) We cannot apply the Divergence Theorem to the whole region within the box, because the vector field $\vec{F}$ is not defined at the origin. However, we can apply the Divergence Theorem to the region, W, between the sphere and the box. Since $\operatorname{div}\vec{F} = 0$ there, the theorem tells us that

$$\int_{\substack{\text{Box}\\(\text{outward})}} \vec{F}\cdot d\vec{A} + \int_{\substack{\text{Sphere}\\(\text{inward})}} \vec{F}\cdot d\vec{A} = \int_W \operatorname{div}\vec{F}\,dV = 0.$$

Therefore, the flux through the box and the sphere are equal if both are oriented outward:

$$\int_{\substack{\text{Box}\\(\text{outward})}} \vec{F}\cdot d\vec{A} = -\int_{\substack{\text{Sphere}\\(\text{inward})}} \vec{F}\cdot d\vec{A} = \int_{\substack{\text{Sphere}\\(\text{outward})}} \vec{F}\cdot d\vec{A} = 4\pi.$$

37. We find $\operatorname{div}\vec{F} = 1+1+1 = 3$. To use the Divergence Theorem, we need to have a closed surface, so we add a circular disk, D, oriented downward. With S representing the hemisphere and W the interior of the closed region, the Divergence Theorem gives

$$\begin{aligned}\int_{\text{Closed surface}} \vec{F}\cdot d\vec{A} = \int_S \vec{F}\cdot d\vec{A} + \int_D \vec{F}\cdot d\vec{A} = \int_W \operatorname{div}\vec{F}\,dV &= 3\cdot\text{ Volume of region}\\ &= 3\left(\frac{1}{2}\cdot\frac{4}{3}\pi 5^3\right) = 250\pi.\end{aligned}$$

Since D is in the plane $z = 0$ and is oriented downward, $d\vec{A} = -\vec{k}\,dA$ on D, giving

$$\text{Flux through disk} = \int_D \vec{F} \cdot d\vec{A} = \int_D \left((x + \cos y)\vec{i} + (y + \sin x)\vec{j} + 3\vec{k}\right) \cdot (-\vec{k}\,dA)$$
$$= -\int_D 3\,dA = -3\pi(5^2) = -75\pi.$$

Thus,

$$\int_S \vec{F} \cdot d\vec{A} = 250\pi - \int_D \vec{F} \cdot d\vec{A} = 250\pi - (-75\pi) = 325\pi.$$

41. **(a)** Since

$$\vec{F} = F_1\vec{i} + F_2\vec{j} + F_3\vec{k} = \frac{x\vec{i} + y\vec{j} + z\vec{k}}{(x^2 + y^2 + z^2)^{a/2}},$$

we have

$$\frac{\partial F_1}{\partial x} = \frac{1}{(x^2 + y^2 + z^2)^{a/2}} - \frac{a}{2} \cdot \frac{x(2x)}{(x^2 + y^2 + z^2)^{(a/2)+1}}$$
$$= \frac{x^2 + y^2 + z^2 - ax^2}{(x^2 + y^2 + z^2)^{(a/2)+1}}.$$

Similarly, calculating $\partial F_2/\partial y$ and $\partial F_3/\partial z$ and adding gives

$$\operatorname{div} \vec{F} = \frac{\partial F_1}{\partial x} + \frac{\partial F_2}{\partial y} + \frac{\partial F_3}{\partial z} = \frac{3(x^2 + y^2 + z^2) - ax^2 - ay^2 - az^2}{(x^2 + y^2 + z^2)^{(a/2)+1}}$$
$$= \frac{3 - a}{(x^2 + y^2 + z^2)^{a/2}}.$$

(b) $\operatorname{div} \vec{F} = 0$ if $a = 3$.

(c) The vector field $\vec{F}$ is radial, and shows no "swirl", so we expect $\operatorname{curl} \vec{F} = \vec{0}$. See Figure 20.6.

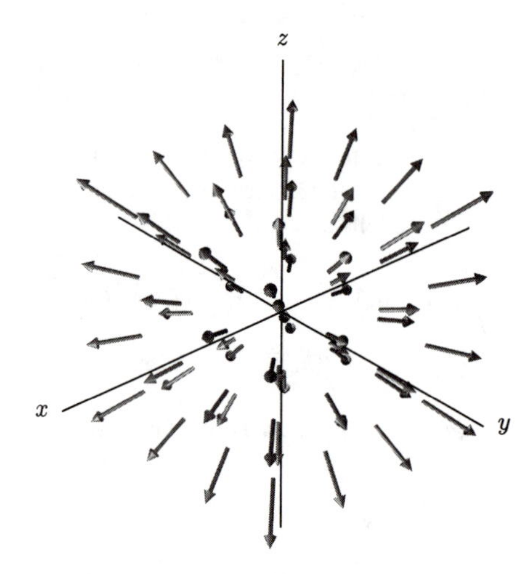

Figure 20.6

45. **(a)** Since $\vec{v} = \operatorname{grad} \phi$ we have

$$\vec{v} = \left(1 + \frac{y^2 - x^2}{(x^2 + y^2)^2}\right)\vec{i} + \frac{-2xy}{(x^2 + y^2)^2}\vec{j}$$

(b) Differentiating the components of $\vec{v}$, we have

$$\operatorname{div} \vec{v} = \frac{\partial}{\partial x}(1 + \frac{y^2 - x^2}{(x^2 + y^2)^2}) + \frac{\partial}{\partial y}(\frac{-2xy}{(x^2 + y^2)^2}) = \frac{2x(x^2 - 3y^2)}{(x^2 + y^2)^3} + \frac{2x(3y^2 - x^2)}{(x^2 + y^2)^3} = 0$$

(c) The vector $\vec{v}$ is tangent to the circle $x^2+y^2=1$, if and only if the dot product of the field on the circle with any radius vector of that circle is zero. Let (x, y) be a point on the circle. We want to show: $\vec{v}\cdot\vec{r}=\vec{v}(x,y)\cdot(x\vec{i}+y\vec{j})=0$. We have:

$$\begin{aligned}\vec{v}(x,y)\cdot(x\vec{i}+y\vec{j}) &= ((1+\frac{y^2-x^2}{(x^2+y^2)^2})\vec{i}+\frac{-2xy}{(x^2+y^2)^2}\vec{j})\cdot(x\vec{i}+y\vec{j})\\ &= x+x\frac{y^2-x^2}{(x^2+y^2)^2}-\frac{2xy^2}{(x^2+y^2)^2}\\ &= \frac{x(x^2+y^2-1)}{x^2+y^2},\end{aligned}$$

but we know that for any point on the circle, $x^2+y^2=1$, thus we have $\vec{v}\cdot\vec{r}=0$. Therefore, the velocity field is tangent to the circle. Consequently, there is no flow through the circle and any water on the outside of the circle must flow around it.

(d)

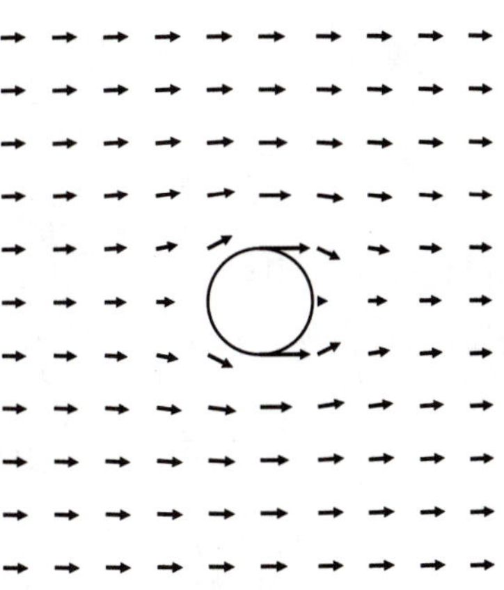

Figure 20.7

49. (a) The path along which we integrate is shown in Figure 20.8.

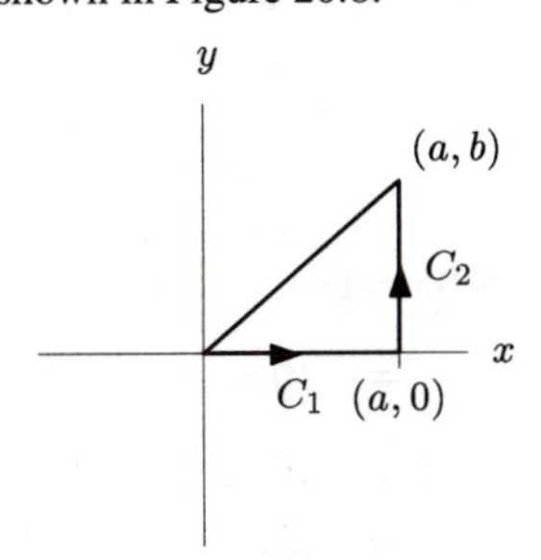

Figure 20.8

The path C_1 is given by

$$C_1:\begin{cases}x=t & 0\le t\le a\\ y=0\end{cases}$$

and path C_2 is given by

$$C_2:\begin{cases}x=a\\ y=t & 0\le t\le b\end{cases}.$$

We integrate along the path $C=C_1+C_2$. Then,

$$\begin{aligned}\int_C \vec{E}\cdot d\vec{r} &= \int_{C_1}\vec{E}\cdot d\vec{r}+\int_{C_2}\vec{E}\cdot d\vec{r}\\ &= \int_0^a 5t^2\vec{j}\cdot\vec{i}\,dt+\int_0^b(10at\vec{i}+(5a^2-5t^2)\vec{j})\cdot\vec{j}\,dt\\ &= 5a^2b-\frac{5}{3}b^3\end{aligned}$$

(b) The path along which we integrate is shown in Figure 20.9.

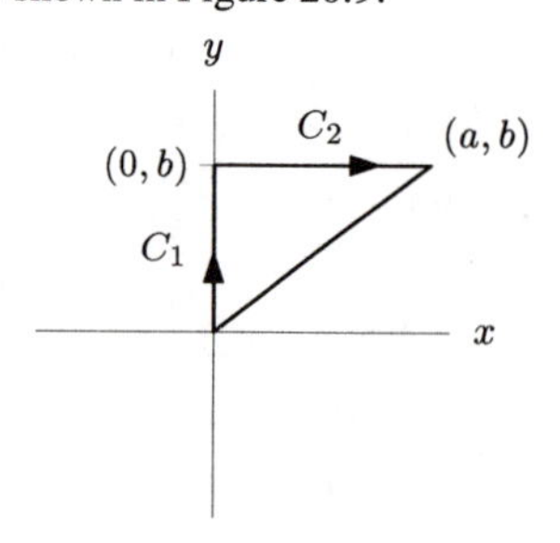

Figure 20.9

The path C_1 is given by

$$C_1 : \begin{cases} x = 0 \\ y = t \quad 0 \le t \le b \end{cases}$$

and Path C_2 is given by

$$C_2 : \begin{cases} x = t \quad 0 \le t \le a \\ y = b \end{cases}.$$

We integrate along the path $C = C_1 + C_2$. Then,

$$\begin{aligned}\int_C \vec{E} \cdot d\vec{r} &= \int_{C_1} \vec{E} \cdot d\vec{r} + \int_{C_2} \vec{E} \cdot d\vec{r} \\ &= \int_0^b -5t^2\vec{j} \cdot \vec{j}\, dt + \int_0^a (10bt\vec{i} + (5t^2 - 5b^2)\vec{j}) \cdot \vec{i}\, dt \\ &= -\frac{5}{3}b^3 + 5a^2b\end{aligned}$$

(c) Notice that the line integrals along both paths in part (a) and (b) are equal. Thus $\vec{E}$ could be path independent. This property is a property of electric fields.

(d) Using the calculations of the parts (a) and (b), with the point (a,b) replaced by (x,y), we see that the electric potential, ϕ, at (x,y) is given by

$$\phi = \frac{5}{3}y^3 - 5x^2y.$$

Then, taking the gradient gives

$$\begin{aligned}\text{grad}\,\phi &= \phi_x\vec{i} + \phi_y\vec{j} \\ &= -10xy\vec{i} + (\frac{5}{3}(3y^2) - 5x^2)\vec{j} \\ &= -10xy\vec{i} + (5y^2 - 5x^2)\vec{j} \\ &= -\vec{E}.\end{aligned}$$

Thus, we have confirmed that $\vec{E} = -\,\text{grad}\,\phi$.

CAS Challenge Problems

53. (a) Let W be the region enclosed by the sphere. We have div $\vec{F} = 2ax+bz+2cy+p+q$ so by the Divergence Theorem $\int_S \vec{F} \cdot d\vec{A} = \int_W (2ax+bz+2cy+p+q)\,dV$. Now $\int_W x\,dV = \int_W y\,dV = \int_W z\,dV = 0$, because W is symmetric about the origin and x, y, and z are odd functions. So $\int_W (2ax+bz+2cy+p+q)\,dV = \int_B (p+q)dV = \frac{4(p+q)\pi R^3}{3}$.

(b) Using spherical coordinates, we calculate the flux integral directly as

$$\begin{aligned}&\int_0^{2\pi}\int_0^{\pi} ((bR^2\cos(\theta)\cos(\phi)\sin(\phi) + aR^2\cos(\theta)^2\sin(\phi)^2)\vec{i} \\ &+(pR\sin(\theta)\sin(\phi) + cR^2\sin(\theta)^2\sin(\phi)^2)\vec{j} + (qR\cos(\phi) \\ &+rR^3\cos(\theta)^3\sin(\phi)^3)\vec{k}) \cdot (\sin\phi\cos\theta\vec{i} + \sin\phi\sin\theta\vec{j} + \cos\phi\vec{k})R^2\sin\phi\, d\phi d\theta = \frac{4(p+q)\pi R^3}{3}.\end{aligned}$$

Rather than entering this integral directly into your CAS, it is better to define the vector field and parameterization separately and enter the formula for flux integral through a sphere.

CHECK YOUR UNDERSTANDING

1. True. By Stokes' Theorem, the circulation of $\vec{F}$ around C is the flux of curl $\vec{F}$ through the flat disc S in the xy-plane enclosed by the circle. An area element for S is $d\vec{A} = \pm\vec{k}\, dA$, where the sign depends on the orientation of the circle. Since curl $\vec{F}$ is perpendicular to the z-axis, curl $\vec{F} \cdot d\vec{A} = \pm(\text{curl}\,\vec{F} \cdot \vec{k})dA = 0$, so the flux of curl $\vec{F}$ through S is zero, hence the circulation of $\vec{F}$ around C is zero.

5. True. div $\vec{F}$ is a scalar whose value depends on the point at which it is calculated.

9. False. As a counterexample, consider $\vec{F} = 2x\vec{i} + 2y\vec{j} + 2z\vec{k}$. Then $\vec{F} = \text{grad}(x^2+y^2+z^2)$, and div $\vec{F} = 2+2+2 \neq 0$.

13. False. As a counterexample, note that $\vec{F} = \vec{i}$ and $\vec{G} = \vec{j}$ both have divergence zero, but are not the same vector fields.

17. True. The Divergence theorem says that $\int_W \text{div}\,\vec{F}\, dV = \int_S \vec{F} \cdot d\vec{A}$, where S is the outward oriented boundary of W. In this case, the boundary of W consists of the surfaces S_1 and S_2. To give this boundary surface a consistent outward orientation, we use a normal vector on S_1 that points towards the origin, and a normal on S_2 that points away from the origin. Thus $\int_W \text{div}\,\vec{F}\, dV = \int_{S_2} \vec{F} \cdot d\vec{A} + \int_{S_1} \vec{F} \cdot d\vec{A}$, with S_2 oriented outward and S_1 oriented inward. Reversing the orientation on S_1 so that both spheres are oriented outward yields $\int_W \text{div}\,\vec{F}\, dV = \int_{S_2} \vec{F} \cdot d\vec{A} - \int_{S_1} \vec{F} \cdot d\vec{A}$.

21. False. The left-hand side of the equation, $\text{div}(\text{grad}\, f)$, is a scalar function and the right hand side, $\text{grad}(\text{div}\, F)$, is a vector. There cannot be an equality between a scalar and a vector.

25. True. Writing $\vec{F} = F_1\vec{i} + F_2\vec{j} + F_3\vec{k}$ and $\vec{G} = G_1\vec{i} + G_2\vec{j} + G_3\vec{k}$, we have $\vec{F} + \vec{G} = (F_1+G_1)\vec{i} + (F_2+G_2)\vec{j} + (F_3+G_3)\vec{k}$. Then the $\vec{i}$ component of $\text{curl}(\vec{F} + \vec{G})$ is

$$\frac{\partial(F_3+G_3)}{\partial y} - \frac{\partial(F_2+G_2)}{\partial z} = \frac{\partial F_3}{\partial y} - \frac{\partial F_2}{\partial z} + \frac{\partial G_3}{\partial y} - \frac{\partial G_2}{\partial z}$$

which is the $\vec{i}$ component of curl$\vec{F}$ plus the $\vec{i}$ component of curl$\vec{G}$. The $\vec{j}$ and $\vec{k}$ components work out in a similar manner.

29. False. For example, take $\vec{F} = z\vec{i} + x\vec{j}$. Then curl$\vec{F} = \vec{j} + \vec{k}$, which is not perpendicular to $\vec{F}$, since $(z\vec{i} + x\vec{j}) \cdot (\vec{j} + \vec{k}) = x \neq 0$.

33. False. The curl needs to be in the flux integral, not the line integral, for a correct statement of Stokes' theorem: $\int_C \vec{F} \cdot d\vec{r} = \int_S \text{curl}\,\vec{F} \cdot d\vec{A}$.

37. True. Let S be the rectangular region inside C, oriented by the right hand rule. By Stokes' theorem, $\int_C \vec{F} \cdot d\vec{r} = \int_S \text{curl}\vec{F} \cdot d\vec{A} = 0$.

41. False. The condition that $\int_S \text{curl}\vec{F} \cdot d\vec{A} = 0$ implies, by Stokes' theorem, that $\int_C \vec{F} \cdot d\vec{r} = 0$. However, $\vec{F}$ need not be a gradient field for this to occur. For example, let $\vec{F} = x\vec{k}$, and let S be the upper unit hemisphere $x^2+y^2+z^2 = 1$, $z \geq 0$ oriented upward. Then C is the circle $x^2 + y^2 = 1$, $z = 0$ oriented counterclockwise when viewed from above. The line integral $\int_C x\vec{k} \cdot d\vec{r} = 0$, since the field $\vec{F}$ is everywhere perpendicular to C. The curl of $\vec{F}$ is the constant field $-\vec{j}$, so $\vec{F}$ is not a gradient field. Yet we have $\int_S -\vec{j} \cdot d\vec{A} = 0$, since the constant field $-\vec{j}$ flows in, and then out of the hemisphere S.

APPENDIX

Solutions for Section A

1. The graph is

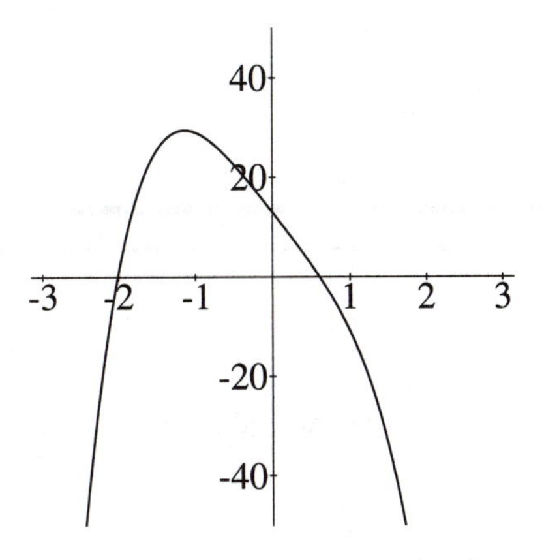

(a) The range appears to be $y \leq 30$.
(b) The function has two zeros.

5. The largest root is at about 2.5.

9. Using a graphing calculator, we see that when x is around 0.45, the graphs intersect.

13. **(a)** Only one real zero, at about $x = -1.15$.
(b) Three real zeros: at $x = 1$, and at about $x = 1.41$ and $x = -1.41$.

17. **(a)** Since f is continuous, there must be one zero between $\theta = 1.4$ and $\theta = 1.6$, and another between $\theta = 1.6$ and $\theta = 1.8$. These are the only clear cases. We might also want to investigate the interval $0.6 \leq \theta \leq 0.8$ since $f(\theta)$ takes on values close to zero on at least part of this interval. Now, $\theta = 0.7$ is in this interval, and $f(0.7) = -0.01 < 0$, so f changes sign twice between $\theta = 0.6$ and $\theta = 0.8$ and hence has two zeros on this interval (assuming f is not *really* wiggly here, which it's not). There are a total of 4 zeros.
(b) As an example, we find the zero of f between $\theta = 0.6$ and $\theta = 0.7$. $f(0.65)$ is positive; $f(0.66)$ is negative. So this zero is contained in $[0.65, 0.66]$. The other zeros are contained in the intervals $[0.72, 0.73]$, $[1.43, 1.44]$, and $[1.7, 1.71]$.
(c) You've found all the zeros. A picture will confirm this; see Figure A.1.

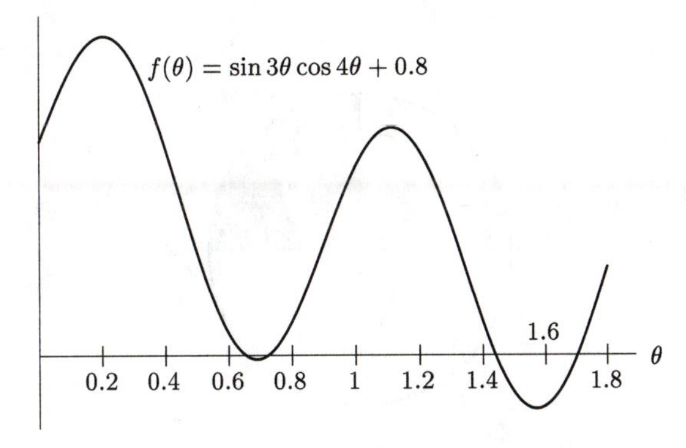

Figure A.1

21.

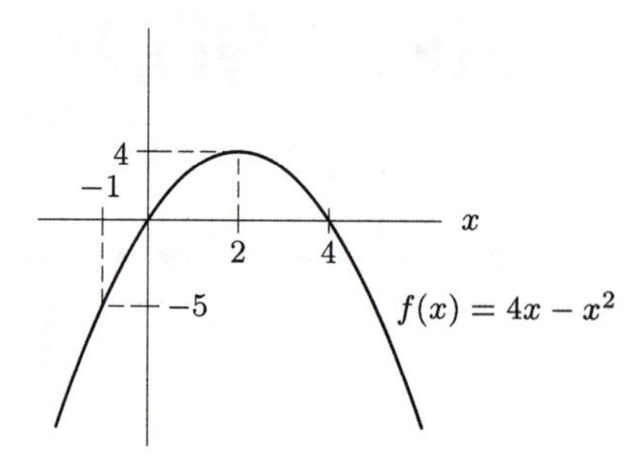

Bounded and $-5 \leq f(x) \leq 4$.

Solutions for Section B

1. $r = \sqrt{1^2 + 0^2} = 1, \quad \theta = 0.$

5. $r = \sqrt{(-3)^2 + (-3)^2} = 4.2.$
$\tan\theta = (-3/-3) = 1$. Since the point is in the third quadrant, $\theta = 5\pi/4$.

9. (1,0)

13. $(\frac{5\sqrt{3}}{2}, -\frac{5}{2})$

17. The graph is a circle of radius 2 centered at the origin. See Figure B.2.

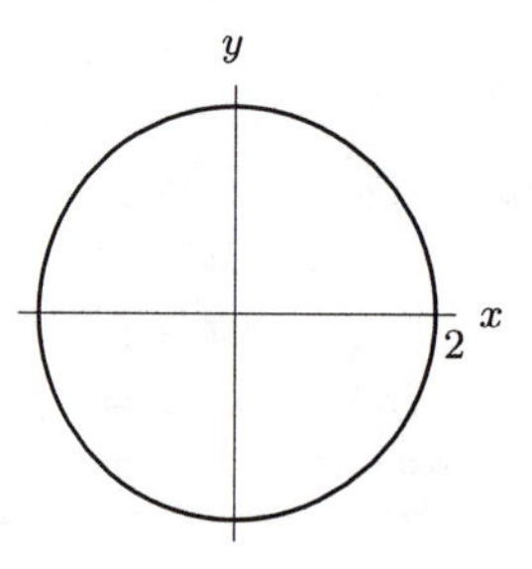

Figure B.2

21. The condition that $r \leq 2$ tells us that the region is inside a circle of radius 2 centered at the origin. The second condition $0 \leq \theta \leq \pi/2$ tells us that the points must be in the first quadrant. Thus, the region consists of the quarter of the circle in the first quadrant, as shown in Figure B.3.

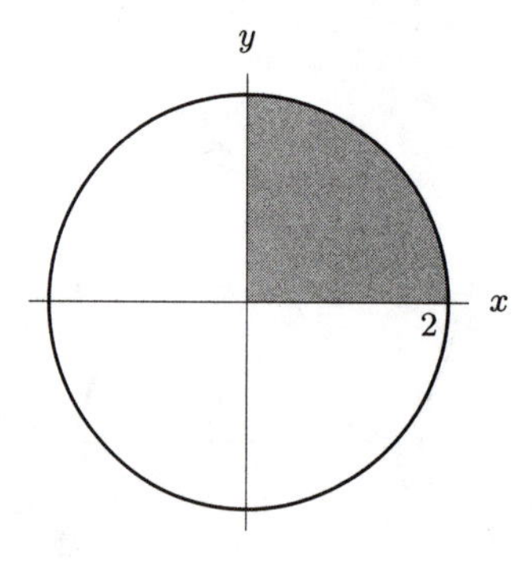

Figure B.3

25. The region consists of the portion of a circle of radius 1 centered at the origin that is between an angle of $\theta = 0$ and an angle of $\theta = \pi/4$, Therefore, the region is defined by $r \leq 1$ and $0 \leq \theta \leq \pi/4$.

29. Putting $\theta = \pi/3$ into $\tan\theta = y/x$ gives $\sqrt{3} = y/x$, or $y = \sqrt{3}x$. This is a line through the origin of slope $\sqrt{3}$. See Figure B.4.

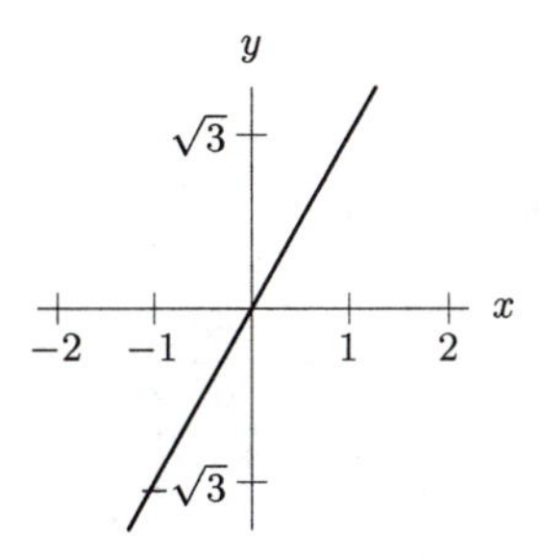

Figure B.4

33. For $r = \theta/10$, the radius r increases as the angle θ winds around the origin, so this is a spiral. See Figure B.5.

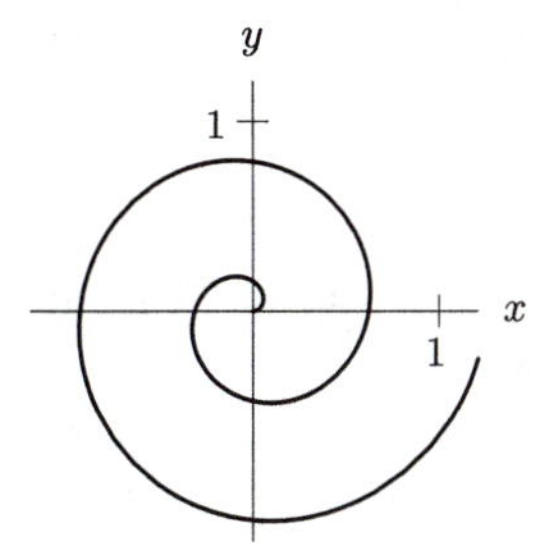

Figure B.5

Solutions for Section C

1. $2e^{i\pi/2}$

5. $0e^{i\theta}$, for any θ.

9. $-3 - 4i$

13. $\frac{1}{4} - \frac{9i}{8}$

17. $5^3(\cos\frac{3\pi}{2} + i\sin\frac{3\pi}{2}) = -125i$

21. One value of $\sqrt[3]{i}$ is $\sqrt[3]{e^{i\frac{\pi}{2}}} = (e^{i\frac{\pi}{2}})^{\frac{1}{3}} = e^{i\frac{\pi}{6}} = \cos\frac{\pi}{6} + i\sin\frac{\pi}{6} = \frac{\sqrt{3}}{2} + \frac{i}{2}$

25. One value of $(-4+4i)^{2/3}$ is $[\sqrt{32}e^{(i3\pi/4)}]^{(2/3)} = (\sqrt{32})^{2/3}e^{(i\pi/2)} = 2^{5/3}\cos\frac{\pi}{2} + i2^{5/3}\sin\frac{\pi}{2} = 2i\sqrt[3]{4}$

29. We have

$$
\begin{aligned}
i^{-1} &= \frac{1}{i} = \frac{1}{i}\cdot\frac{i}{i} = -i,\\
i^{-2} &= \frac{1}{i^2} = -1,\\
i^{-3} &= \frac{1}{i^3} = \frac{1}{-i}\cdot\frac{i}{i} = i,\\
i^{-4} &= \frac{1}{i^4} = 1.
\end{aligned}
$$

The pattern is

$$i^n = \begin{cases} -i & n = -1, -5, -9, \cdots \\ -1 & n = -2, -6, -10, \cdots \\ i & n = -3, -7, -11, \cdots \\ 1 & n = -4, -8, -12, \cdots . \end{cases}$$

Since 36 is a multiple of 4, we know $i^{-36} = 1$.
Since $41 = 4 \cdot 10 + 1$, we know $i^{-41} = -i$.

33. To confirm that $z = \dfrac{a+bi}{c+di}$, we calculate the product

$$\begin{aligned} z(c+di) &= \left(\frac{ac+bd}{c^2+d^2} = \frac{bc-ad}{c^2+d^2}i\right)(c+di) \\ &= \frac{ac^2+bcd-bcd+ad^2+(bc^2-acd+acd+bd^2)i}{c^2+d^2} \\ &= \frac{a(c^2+d^2)+b(c^2+d^2)i}{c^2+d^2} = a+bi. \end{aligned}$$

37. True, since $\sqrt{a}$ is real for all $a \geq 0$.

41. True. We can write any nonzero complex number z as $re^{i\beta}$, where r and β are real numbers with $r > 0$. Since $r > 0$, we can write $r = e^c$ for some real number c. Therefore, $z = re^{i\beta} = e^c e^{i\beta} = e^{c+i\beta} = e^w$ where $w = c + i\beta$ is a complex number.

45. Using Euler's formula, we have:

$$e^{i(2\theta)} = \cos 2\theta + i \sin 2\theta$$

On the other hand,

$$e^{i(2\theta)} = \left(e^{i\theta}\right)^2 = (\cos\theta + i\sin\theta)^2 = (\cos^2\theta - \sin^2\theta) + i(2\cos\theta\sin\theta)$$

Equating real parts, we find

$$\cos 2\theta = \cos^2\theta - \sin^2\theta.$$

49. Replacing θ by $(x+y)$ in the formula for $\sin\theta$:

$$\begin{aligned} \sin(x+y) &= \frac{1}{2i}\left(e^{i(x+y)} - e^{-i(x+y)}\right) = \frac{1}{2i}\left(e^{ix}e^{iy} - e^{-ix}e^{-iy}\right) \\ &= \frac{1}{2i}\left((\cos x + i\sin x)(\cos y + i\sin y) - (\cos(-x) + i\sin(-x))(\cos(-y) + i\sin(-y))\right) \\ &= \frac{1}{2i}\left((\cos x + i\sin x)(\cos y + i\sin y) - (\cos x - i\sin x)(\cos y - i\sin y)\right) \\ &= \sin x\cos y + \cos x\sin y. \end{aligned}$$

Solutions for Section D

1. **(a)** $f'(x) = 3x^2 + 6x + 3 = 3(x+1)^2$. Thus $f'(x) > 0$ everywhere except at $x = -1$, so it is increasing everywhere except perhaps at $x = -1$. The function is in fact increasing at $x = -1$ since $f(x) > f(-1)$ for $x > -1$, and $f(x) < f(-1)$ for $x < -1$.

(b) The original equation can have at most one root, since it can only pass through the x-axis once if it never decreases. It must have one root, since $f(0) = -6$ and $f(1) = 1$.

(c) The root is in the interval $[0, 1]$, since $f(0) < 0 < f(1)$.

(d) Let $x_0 = 1$.

$$\begin{aligned} x_0 &= 1 \\ x_1 &= 1 - \frac{f(1)}{f'(1)} = 1 - \frac{1}{12} = \frac{11}{12} \approx 0.917 \\ x_2 &= \frac{11}{12} - \frac{f\left(\frac{11}{12}\right)}{f'\left(\frac{11}{12}\right)} \approx 0.913 \\ x_3 &= 0.913 - \frac{f(0.913)}{f'(0.913)} \approx 0.913. \end{aligned}$$

Since the digits repeat, they should be accurate. Thus $x \approx 0.913$.

5. Let $f(x) = \sin x - 1 + x$; we want to find all zeros of f, because $f(x) = 0$ implies $\sin x = 1 - x$. Graphing $\sin x$ and $1 - x$ in Figure D.6, we see that $f(x)$ has one solution at $x \approx \frac{1}{2}$.

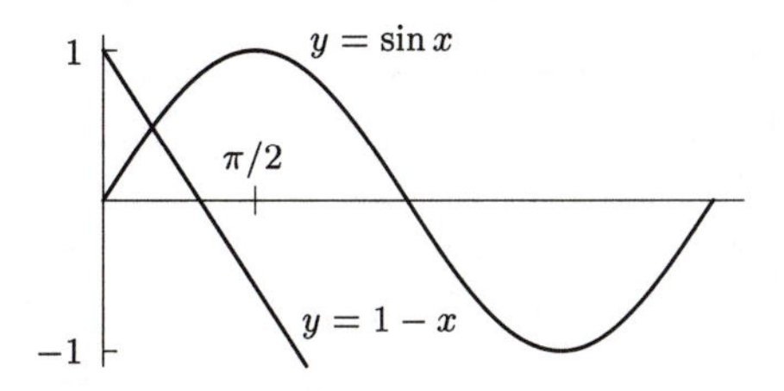

Figure D.6

Letting $x_0 = 0.5$, and using Newton's method, we have $f'(x) = \cos x + 1$, so that

$$x_1 = 0.5 - \frac{\sin(0.5) - 1 + 0.5}{\cos(0.5) + 1} \approx 0.511,$$

$$x_2 = 0.511 - \frac{\sin(0.511) - 1 + 0.511}{\cos(0.511) + 1} \approx 0.511.$$

Thus $\sin x = 1 - x$ has one solution at $x \approx 0.511$.

9. Let $f(x) = \ln x - \frac{1}{x}$, so $f'(x) = \frac{1}{x} + \frac{1}{x^2}$.
Now use Newton's method with an initial guess of $x_0 = 2$.

$$\begin{aligned} x_1 &= 2 - \frac{\ln 2 - \frac{1}{2}}{\frac{1}{2} + \frac{1}{4}} \approx 1.7425, \\ x_2 &\approx 1.763, \\ x_3 &\approx 1.763. \end{aligned}$$

Thus $x \approx 1.763$ is a solution. Since $f'(x) > 0$ for positive x, f is increasing: it must be the only solution.